FIFTY MAPS

AND THE STORIES THEY TELL

Arctic Circle
OKHOTSK BASIN
ALEUTIAN BASIN
MAURY DEEP
JAPAN BASIN
Deepest authentic sounding known
TUSCARORA DEEP
MURRAY DEEP
CLOVER DEEP
TANNER DEEP
HAWAIAN PLATEAU
Tropic of Cancer
WYMAN DEEP
AMMEN DEEP
RENARD DEEP
CALIFOR PLATE
LADRONE PLATEAU
BROOKE DEEP
BELKNAP DEEP
sounding of "Challenger"
MARSHALL PLATEAU
GREY DEEP
LINE PLATEAU
CAMPBELL DEEP
MILLER DEEP
Equator
150
160
170
180
170
160
150
140
130
120
SOLOMON ISLANDS PLATEAU
HILGARD DEEP
MARQUESAS PLATEAU
CARPENTER BASIN
AGASSIZ BASIN
FIJI PLATEAU
OLDHAM DEEP
PAUMOTU PLATEAU
Tropic of Capricorn
PLAT
GAZELLE BASIN
NEW ZEALA
ALDRICH DEEP

FIFTY MAPS

AND THE STORIES THEY TELL

Jerry Brotton
and Nick Millea

BODLEIAN
LIBRARY
PUBLISHING

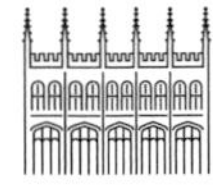

First published in 2019 by the Bodleian Library
Broad Street, Oxford OX1 3BG
www.bodleianshop.co.uk

2nd impression 2022

ISBN: 978 1 85124 523 9

Cover design by Dot Little at the Bodleian Library
Designed and typeset by Laura Parker in Trivia Sans in 8.8 on 13 point
Printed and bound by Livonia Print, Latvia, on 150 gsm Magno matt paper

British Library Catalogue in Publishing Data
A CIP record of this publication is available from the British Library

CONTENTS

गंगानदी
महाहिमवंत पर्वत
हरिकांतानदी
गंधापाती

INTRODUCTION

Every map tells a story, and *Fifty Maps* is a celebration of the sheer variety of those stories, told by travellers, sailors, merchants, pilgrims and many others. These stories in turn create their own. Most people regard maps as route-finding devices, and this aspect of their history is certainly told in some detail in this book. Over time, however, maps have been concerned with so much more than simply getting us from A to B. As we will see, this function of maps only occupies a small and quite recent part of the longer story of map-making.

It was only in Europe in the nineteenth century, when the term 'cartography' was first used to describe the scientific study and practice of map-making, that maps became established as transparent and accurate devices enabling us to get from one place to another. Since then, advances in the organization of mapping such as the Ordnance Survey – Great Britain's national map agency – have created the conditions for printed maps and atlases to take their place in all aspects of everyday life, from the national, economic and regional administration of towns, waterways and agriculture, to individuals using them to navigate their way across cities or the countryside. In the process, maps have become ubiquitous, perhaps even more now than ever as they have gone online and we use them

Detail from a Jain map showing the holy mountain of Meru, sixteenth–seventeenth century. MS. Or. Evans-Wentz 1.

in internet searches for everything from town planning to finding the nearest hotel or restaurant.

Many of the maps reproduced in the following pages recount stories of how maps orientate us and enable us to move confidently within our environment. However, in the course of our book we also discover many different roles for maps. Throughout time and across very different cultures we see that maps have been central to how people have told stories about their sacred places, such as Mecca, Mount Meru and Christianity's heaven and Garden of Eden, and about the soul's journey through various states of being in the process of transmigration and rebirth seen in Buddhist and Jain cosmological maps. Other maps called cosmogonies tell the story of the world's beginning – and what better artefact to use to describe creation than a map? We also discover how maps have been used by writers and artists to imagine through graphic visualization alternative places and fantasy worlds. We trace a long and distinguished tradition of such maps, ranging from those illustrating Sir Thomas More's *Utopia* in the early sixteenth century, through Robert Louis Stevenson's nineteenth-century map of Treasure Island, to J.R.R. Tolkien's maps of Middle-earth and C.S. Lewis's map of Narnia from the mid-twentieth century.

Just as writers have been drawn to the power of maps in enabling them to imagine and tell their stories, visual artists have also been fascinated by how a map can combine words and images to create an object that seems to provide a very convincing representation of reality. It is only relatively recently that the creation of maps has been the responsibility of technical draughtsmen using scientific principles. Throughout much of the history of map-making artists were responsible for every aspect of their design, from shape and size to colour. Contemporary artists acknowledge this history while also using maps in their art to ask to what extent a map is ever a true representation of the place it claims to show.

The only map that could exactly reproduce the territory it depicts would be on a scale of 1:1, which would of course be of no use whatsoever. In Lewis Carroll's novel *Sylvie and Bruno Concluded* (1893), one of the characters announces that they 'made a map of the country, on a scale of *a mile to the mile*!' When asked about the map's use the character admits, 'It has never been spread out', and that 'the farmers objected: they said it would cover the whole country, and shut out the sunlight! So we now use the country itself, as its own map, and I assure you it does nearly as well.'[1] Behind Carroll's amusing story is a more serious point about how we require maps to be highly selective about what they include – and what they omit – in managing the vast complexity of the world we inhabit. Such decisions are also central to the process of making art, and we illustrate how largely anonymous European, Arab, Persian, Chinese, Tibetan, Indian, Mesoamerican and Pacific island artists have made a variety of decisions in making maps, using materials, signs, colours and dimensions unique to their cultural beliefs and preoccupations. We also show how contemporary artists including Layla Curtis and Grayson Perry create explicit art maps or use mapping techniques in their works to reflect on the importance of maps to our sense of who we are, by posing the question of *where* we are.

By enlarging our sense of maps in this way, we follow recent academic scholarship in defining maps as 'graphic representations that facilitate a spatial understanding of things, concepts,

1 Lewis Carroll, *Sylvie and Bruno Concluded*, London, 1893, p. 169.

A scale bar from Christopher Saxton, *Atlas of the Counties of England and Wales*. MAP RES 79, fol. 12.

conditions, processes, or events in the human world.'[2] This definition can be applied to each of the fifty maps found in this book. Some of them are described as charts (usually related to maritime navigation), plots (an archaic term for a map, that shares a striking root with the plot of a story), *mappae mundi* (Latin for 'world maps'), or simply drawings. Many are made on paper, but some are on animal skin, some woven in wool, others made from coconut fronds, and some are created from photographic and digital technology and projected onto computer screens. Yet all of them share a common aim, to create a graphic image that enables people to better understand the spaces within which they live, be they 'real' places such as the land, the sea or the nation state; imagined ones such as Utopia; or the afterlife of heaven (or hell).

2 J.B. Harley and David Woodward (eds), *The History of Cartography*, Vol. I: *Cartography in Prehistoric, Ancient and Medieval Europe and the Mediterranean*, Chicago and London, 1987, p. xvi.

Fifty Maps is chronological, covering nearly two millennia, beginning in Alexandria in the second century CE and ending with today's computer-generated maps of current world population growth. It is not, however, intended as a short history of cartography. Rather it celebrates the Bodleian Library's remarkable collection of more than 1.5 million maps, one of the best in the world. It includes some of the rarest and most important maps drawn from across history and cultures, and features remarkable recent cartographic discoveries of maps of England, China and the Arab world, some reproduced here for the first time.

Our focus is as much on the stories each map tells as on its history. We give equal consideration to words and images as being graphic, in acknowledgement of that word's origins. The Greek term *graphē* means drawing *and* writing, and as we will see, most – though not all – maps combine words with images. Although maps are often regarded as being concerned with showing space – as our definition above suggests – they can also describe the passage of time, and in this way, they function somewhat like a story. In the European tradition the words used to describe maps often make this connection with telling stories quite explicit. A 'chart' comes from the Greek *khártēs*, meaning to inscribe. A 'plot', as we have mentioned, is a word shared both by fiction and map-making. A plot is usually defined as the main events of a novel, play or film. But in the medieval and early modern period 'plot' also describes a ground plan, map or sea chart.[3]

The stories we tell begin with the Greek Hellenic world and one of its greatest scholars, Claudius Ptolemy, and his *Geography*. While Ptolemy's world map seems to be shaped exclusively by

3 See 'plot, n.', *OED Online*, www.oed.com.

mathematical and geometrical principles in its graticule (a grid of coordinates) of latitude and longitude, it has silently absorbed a millennium of Greek and Roman travellers' stories and navigational manuals (known as *periploi*). Throughout the course of our book we see how over subsequent centuries Ptolemy's scientific approach comes in and out of focus as other imperatives, such as religion, influence the creation of other kinds of maps. Both Muslim and Christian medieval maps tell a story of the world and its creation in their own theological image, with their concentration on sacred places such as Mecca, Jerusalem and Paradise. Over time fiction writers told different stories of these sacred spaces, including Dante Alighieri, whose enormously influential poem *The Divine Comedy* (*c.*1308–20), with its vivid story of the topography of Christian heaven, hell and purgatory, led to a fashion for mapping the afterlife in sixteenth-century Italy. Others turned to the classical past to fashion stories, such as Sir Thomas More, whose satire on an ideal state, *Utopia* (1516), was also mapped by contemporary artists.

The decorative wooden cover of a fifteenth-century atlas of portolan charts (below), and a diptych of St Mark and St Paul from the same manuscript. MS. Douce 390.

·S·MARCVS·EVG·
·S·PAVLVS·

Maps can tell their own stories, especially when made in a set or cycle that represents an event sequentially. A particularly vivid case in point is a series of charts showing the progress of the Spanish Armada in the summer of 1588, and its famous defeat by the English fleet. The ten charts (so called because they mimic the appearance of navigational sea charts) that were drawn in the aftermath of the English victory offer a blow-by-blow account of the story of naval engagement from the first sighting of the Spanish fleet to its defeat and dispersal eleven days later. Each chart takes approximately a day in the conflict, with the action being shown from left to right, almost like a cartoon strip. The charts appear to have been made to circulate news of the English victory as quickly as possible, and were reproduced in a variety of formats: hand-drawn, engraved and even as a series of tapestries that hung in the House of Lords at the Palace of Westminster before being destroyed by fire in 1834. The graphic impact of such charts would have been as powerful as any written account; they are cartographic reportage, telling a story of a crucial moment in English history.

While many of our other maps, like these, tell stories of the sea, others have equally compelling things to say about the land. At the same time that Tudor England was surviving the threat of Spanish invasion, it was also sponsoring the first national atlas of the English and Welsh counties. Completed by the map-maker Christopher Saxton in 1579, this proved so successful that within a decade some of its maps were being woven as tapestries at vast expense. By the mid-seventeenth century, English county and estate map-making was so developed that it could tell a remarkably detailed story of the annual cycle of agrarian life. Mark Pierce's enormous map of

Agricultural life depicted on Mark Pierce's map of the Laxton and Kneesall estates in Nottinghamshire, 1635. MS. C17:48 (9).

the Laxton and Kneesall estates in Nottinghamshire captures a lost world of agricultural life, with its minute depiction of harvesting, ploughing, shepherding and hunting in an open-field system that is largely intact to this day.

Although the story of the mapping of England features heavily in our book, we are also aware that mapping from cultures often marginalized in the history of cartography has its own stories to tell, even though many of them are very different from those of Western map-making. Just as the Ordnance Survey pursued its epic project of mapping the nation state in the nineteenth century, pilots in the Marshall Islands of the Pacific were using 'stick charts' made from coconut fronds to navigate their way across the region's hundreds of islands and atolls, and the Chukchi people of Siberia were making maps out of sealskins depicting their everyday lives and customs. Jains and Buddhists from the Indian subcontinent were weaving fabulously intricate cosmologies showing the terrestrial and other-worldly realms through which followers of these beliefs were convinced they would transmigrate in their search for release – *nirvana* – from the cycle of rebirth. These two-dimensional representations of potentially infinite three-dimensional worlds within worlds act as guides for believers, who regard them as just as 'real' as an Ordnance Survey map, although both represent very different kinds of journeys and tell very disparate stories, many of which are still difficult to decipher.

While many of the stories found in the following maps may be hard for us to understand due to their historical distance, those produced over the last hundred years tell some more familiar ones. Warfare and maps have gone hand in hand throughout history, and here we show maps of the First and Second World Wars that offer a glimpse into some of the horror, heroism and sacrifice of those conflicts. The maps chronicling the late twentieth century's subsequent descent into the paranoia and secrecy of the Cold War, produced by both British and Soviet military intelligence organizations, tell their own depressing stories that seem more prescient now than ever. It is perhaps in response to this politicization of map-making throughout the course of the global conflicts of the previous century that artists have turned to mapping both to question the ways in which maps draw boundaries and can be used to divide us, and to reveal how they can help to imagine different worlds and alternative ways of being. Two contemporary artists, Layla Curtis and Grayson Perry, offer startling perspectives on the current states we are in, using media as disparate as wool and photomontage.

Our story almost inevitably concludes online, with the rise of geospatial applications as ubiquitous presences on our computers and smartphones. As maps migrate from paper to pixels, we come less to an end and more to a new chapter in the history of map-making. While many decry what they believe is the death of paper maps, our final maps, computer-generated 'cartograms', suggest that what we call a map is changing for ever, but that there is still a place for them to tell powerful stories in graphic form about our changing world. Using computational statistics these cartograms are capable of digesting 'big data' to assess election results or offer striking visualizations of population density and distribution. They tell us that however the story of the twenty-first century unfolds, maps will undoubtedly play their part.

Overleaf
Newcastle and Gateshead, shown as an island, detail from Layla Curtis, *NewcastleGateshead*, digital printed collage, 2005.

Dudley
DUDLEY
KARLANTJIPA
KARLANTJIPA NORTH
ALICE SPRINGS - DARWIN
Under Construction
DESERT
WILSON
Shields
NEWCASTLE
North Newcastle
Wallsend
Sheepscot Reversing Falls
Newcastle
Salt Bay
Glidden Pt
Little Pt
Red Beach
Bald Pt
Woody Bay
Skirmish Pt
South Newcastle
Walkers Corners
MINES
Newcastle
Cottonwood Pond
Saline
Rose Hill Rd.
New Castel
Coal City
Shields
North East
Coal Junction Station
NORTH EAST
Shields Dr.
Newcastle Airport
Port Hunter
Newcastle
Raymond Terrace
Williamtown
Grahamstown Lake
Lemon
Newcastle-under-Lyme
Chesterton
Audley
PACIFIC
Carlisle I.
Berwick I.
Little Barrow Point
New Castle Island
Pauline Island
Newcastle
Prudhoe Is
NORTHUMBERLAND ISLANDS
Knight Is
Middle Island
Cape Northumberland
WISCONSIN CENTRAL
Lake

T H U M B E R L A N D
S T R A I T
G A T E S H E A D
CB-29/67F
I S L A N D
67
100
Nisbett
Settlement
Beachlands
Estate
Camp
River
Ruin
Nisbet
Hotel
Newcastle
Pottery
Primrose
Minersville
901
209
Coal Run
61
Cedar
Sch
460
Shields
272
47
NORTHEAST
RD
MECHANICS
Ruin
HICK'S COVE
Hog
Point
Penneshaw
Eastern
Cove
Darlington
Shearer
Point Rd
Tyneham
Cape
Cape
CAPE HART
Broad Bench

1

CLAUDIUS PTOLEMY'S CLASSICAL WORLD MAP

Claudius Ptolemy is regarded as the 'father of geography'. His treatise on map-making, *Geography*, was written in Alexandria c.150 CE and set the standard for subsequent scientific map-making. Drawing on centuries of Graeco-Roman geometry and mathematics, it proposed two projections for drawing maps using a graticule (a grid of coordinates) of latitude and longitude; this illustration shows the second projection. The text listed 8,000 places within the classical world from Scandinavia in the north and Libya in the south to Korea in the east and the Canary Islands in the west. From the late fourteenth century Renaissance map-makers and navigators began to rediscover Ptolemy and this printed edition drawing on his data was published in 1486.

World map in Claudius Ptolemy, *Geographia*, 1486. Arch. B b.19, fol. Map 1

AQUILO VEL BOREAS
CECIAS APELIOTES
SUBSOLANUS
ASIA
MARE INDICUM
PRASODUM
MARE
EURNOTUS
WLTURNUS EURUS

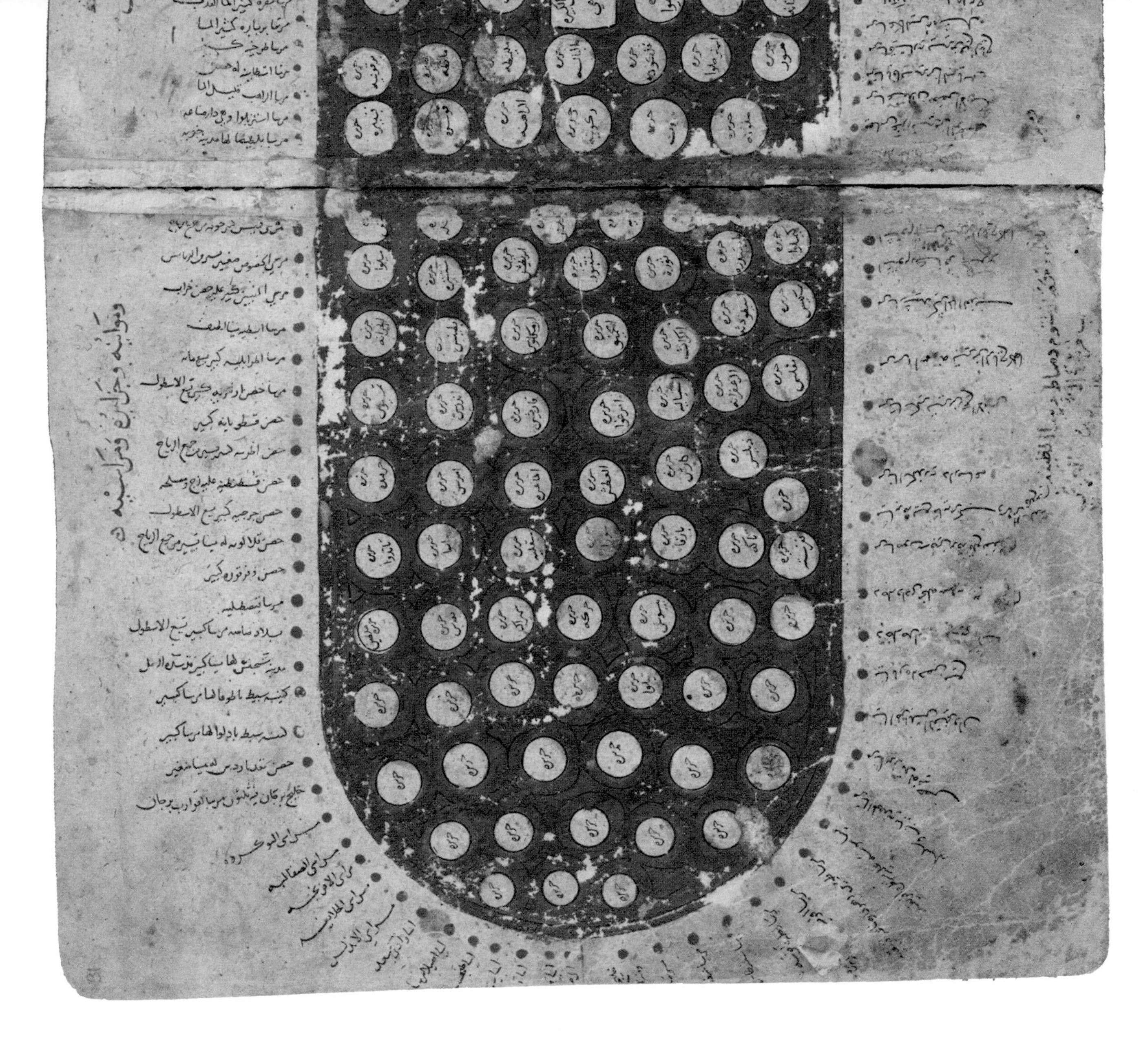

وسواحله وجزائره ومراسيه

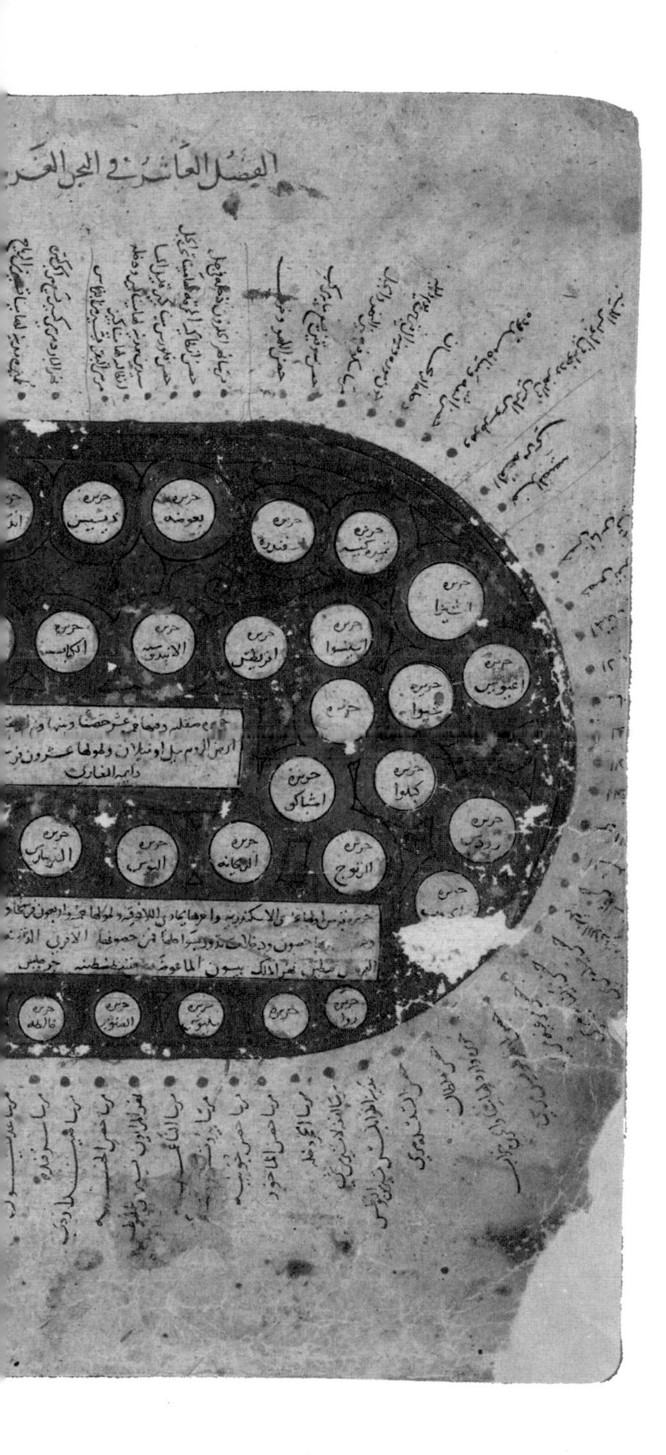

2

A FATIMID MAP OF THE MEDITERRANEAN

This unique map of the Mediterranean comes from an anonymous cosmographical treatise, the *Kitāb Gharā'ib al-funūn* ('The Book of Curiosities of the Sciences and Marvels for the Eyes'), written between 1020 and 1050, under the Egyptian Fatimid dynasty, and which survives in this late twelfth- or early thirteenth-century copy. The map shows 118 islands and 121 harbours, and is oriented to the north, with the Strait of Gibraltar on the far left, Constantinople at top left and Egypt and Libya at the bottom. This is a maritime map that draws its information from the written navigational manuals employed prior to the use of the compass.

Map of the Mediterranean, *The Book of Curiosities*, late twelfth or early thirteenth century. MS. Arab. C. 90, fols 30b–31a

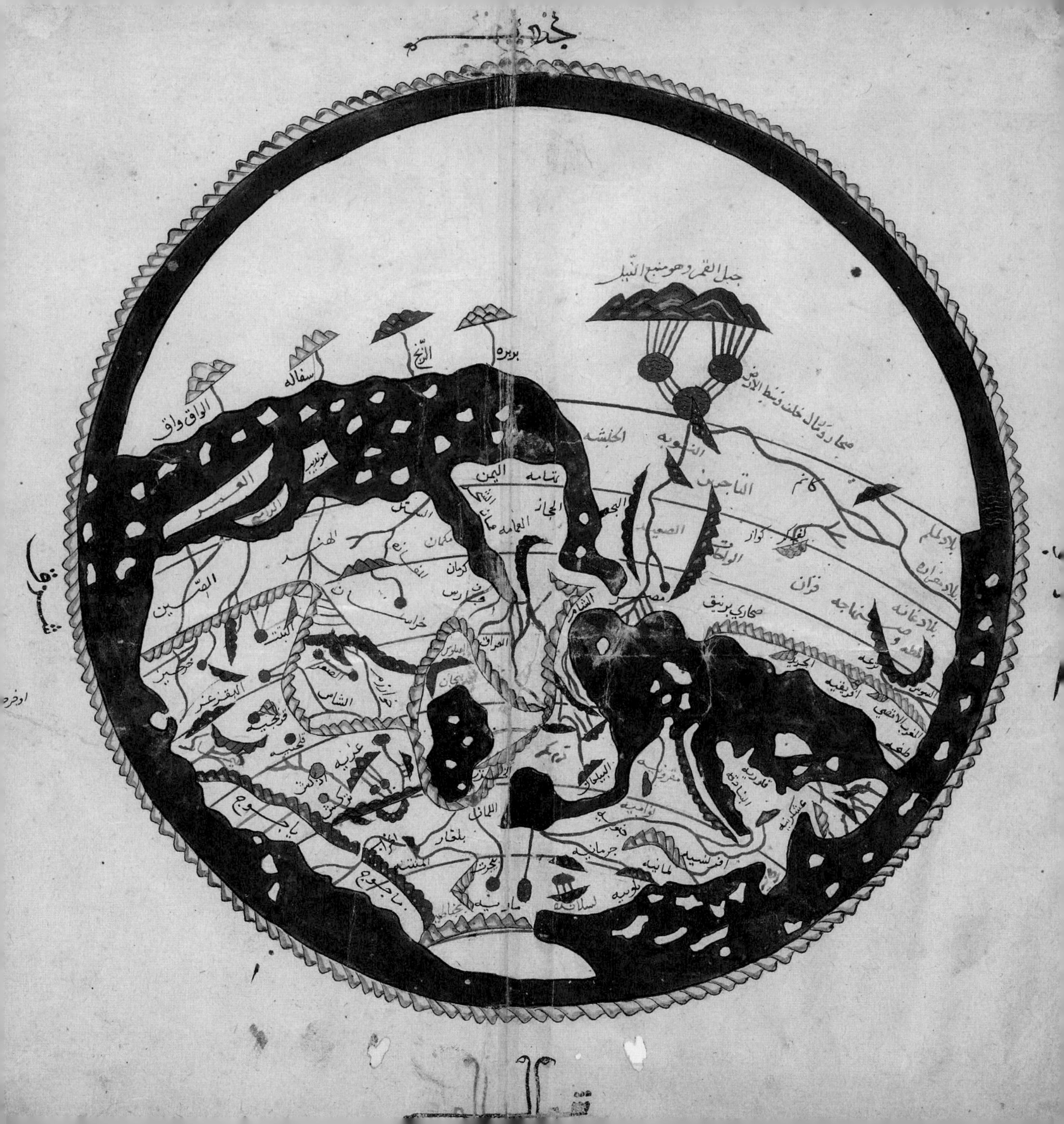

جنوب
جبل القمر وهو منبع النيل
بربره
الزنج
سفاله
الواق واق
الحبشه
النوبه
كانم
اليمن
تهامه
الحجاز
مصر
الشام
العراق
فارس
كرمان
خراسان
الهند
الصين
صحاري برنيق
افريقيه
المغرب الاقصى
ياجوج
ماجوج
بلغار
جرمانيه

3

A MEDIEVAL MUSLIM WORLD MAP

One of the greatest works of medieval map-making was produced in Sicily in the twelfth century by the Muslim scholar al-Sharīf al-Idrīsī. His book the *Nuzhat al-mushtāq* ('Entertainment for He Who Longs to Travel the World') of 1154, one of the most comprehensive descriptions of the inhabited world made since classical times, was commissioned by Sicily's Norman ruler King Roger II. It begins with this circular world map, showing south at the top, with Muslim North Africa and Arabia particularly prominent, as are the mountains believed to be the source of the Nile. It draws on Islamic cosmology and geography but is largely free of theology.

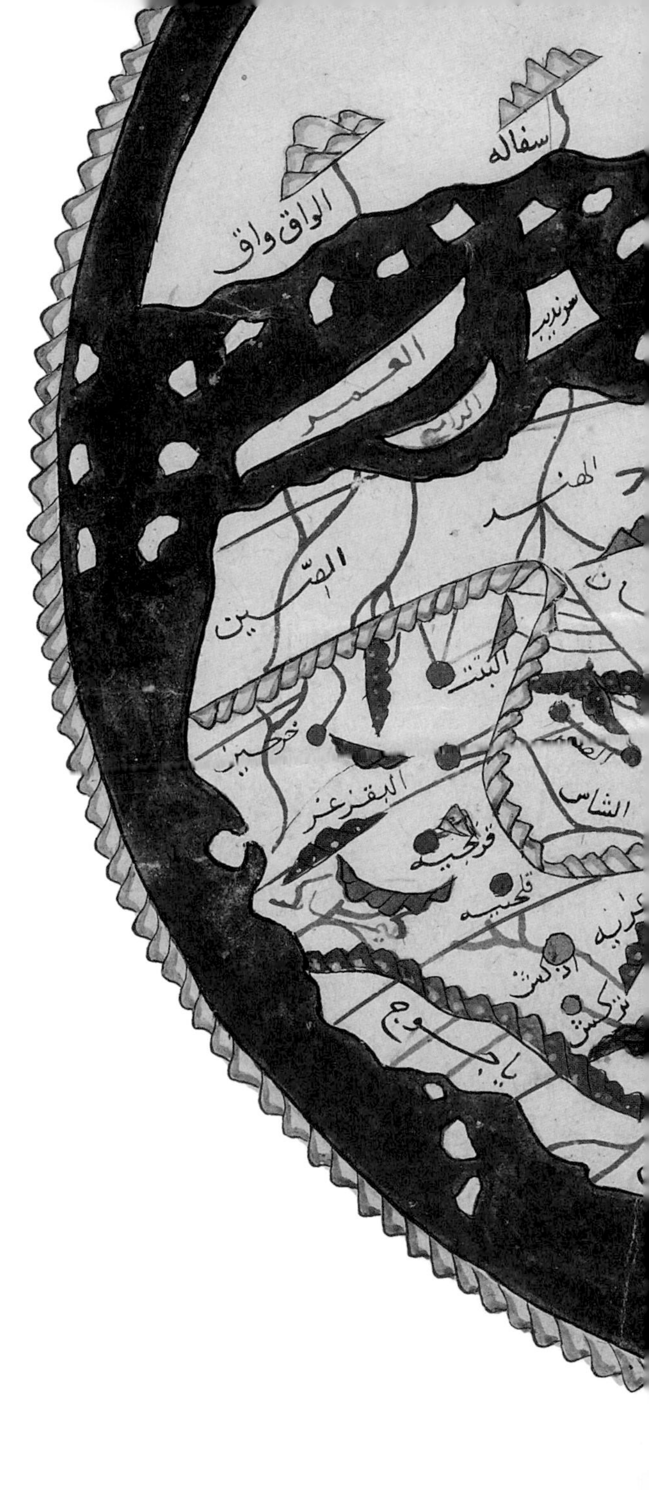

Al-Sharīf al-Idrīsī, circular world map from the *Entertainment*, 1154. MS. Pocock 375, fols 3b–4a

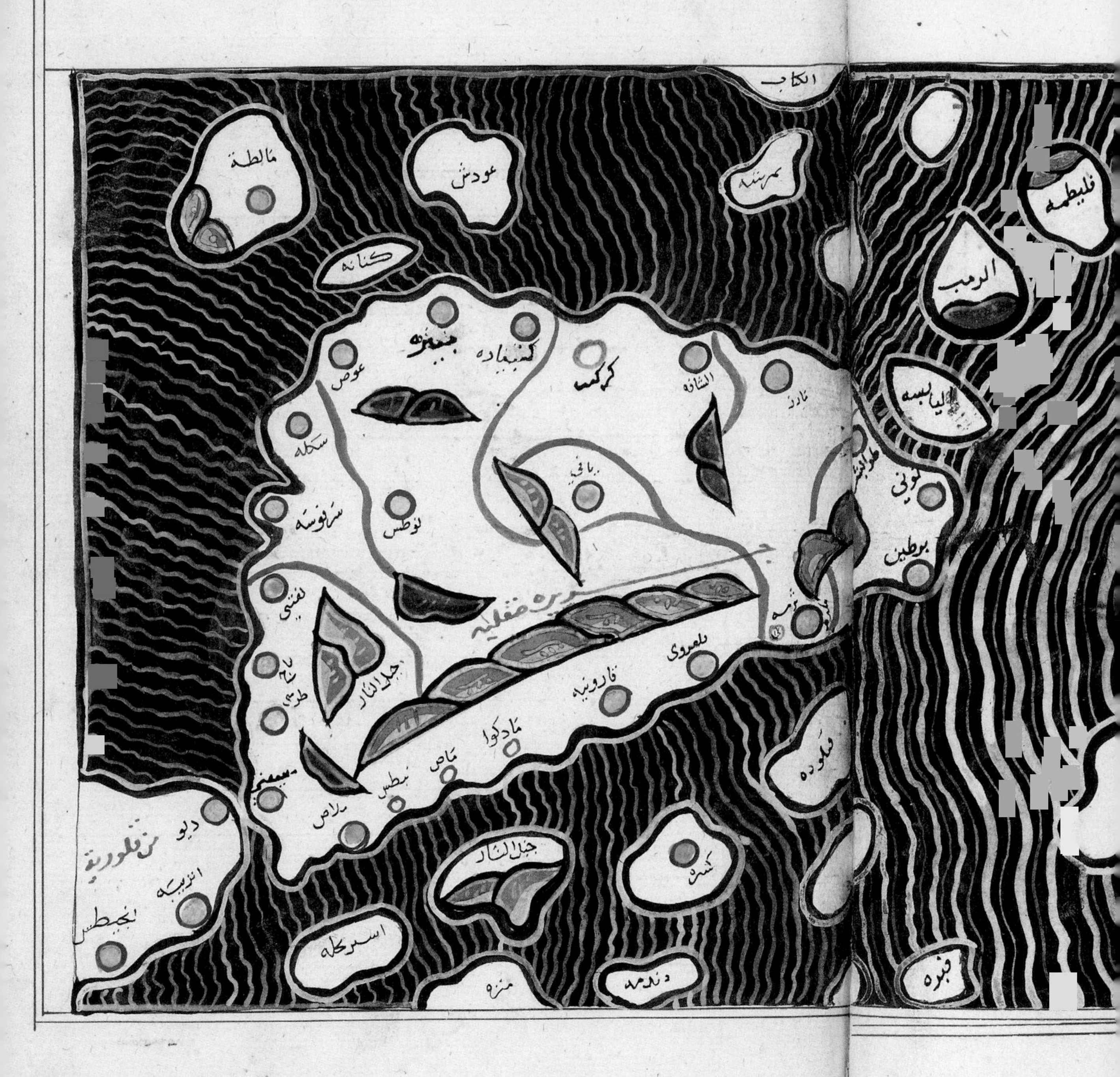
الكاب
مالطة
عودش
كنانه
الرمب
اليابسه
جزيرة صقلية
جبل النار
سرقوسه
لوطس
عوص
سكله
جبل النار
اسركله

Idrīsī's book contained seventy highly detailed regional maps, including a map of Sicily (shown opposite), the home of Roger II, and where Idrīsī worked for many years. South is at the top, with the tip of Italy at bottom left. For its time the map is extremely accurate, capturing the roughly triangular shape of the island, with its main cities and ports labelled by coloured circles, including Roger's imperial capital of Palermo at bottom right. The king's patronage clearly had some influence on the map-making: Idrīsī's Sicily is nearly four times the size of Sardinia (far right), which is in reality only slightly smaller.

Al-Sharīf al-Idrīsī, map of Sicily from the *Entertainment*, 1154. MS. Pococke 375, fols 187b–188a

4

PARADISE FOUND

These two exquisite maps with circumferences of just 60mm are the work of the Italian statesman and philosopher Brunetto Latini. They illustrate his *Livre du Trésor* ('Book of Treasures'), one of the first European encyclopedias. Above is a typical medieval T-O map (see p. 33), showing the East ('Orient') at the top, with the West ('Occident') beneath, split into Europe to the left and Africa to the right. The map below is even more striking: it shows the earthly paradise ('paradis terestre') in the north ('septemtrion') and the Pillars of Hercules (Gibraltar) to the south. On Latini's map, Paradise *is* on Earth.

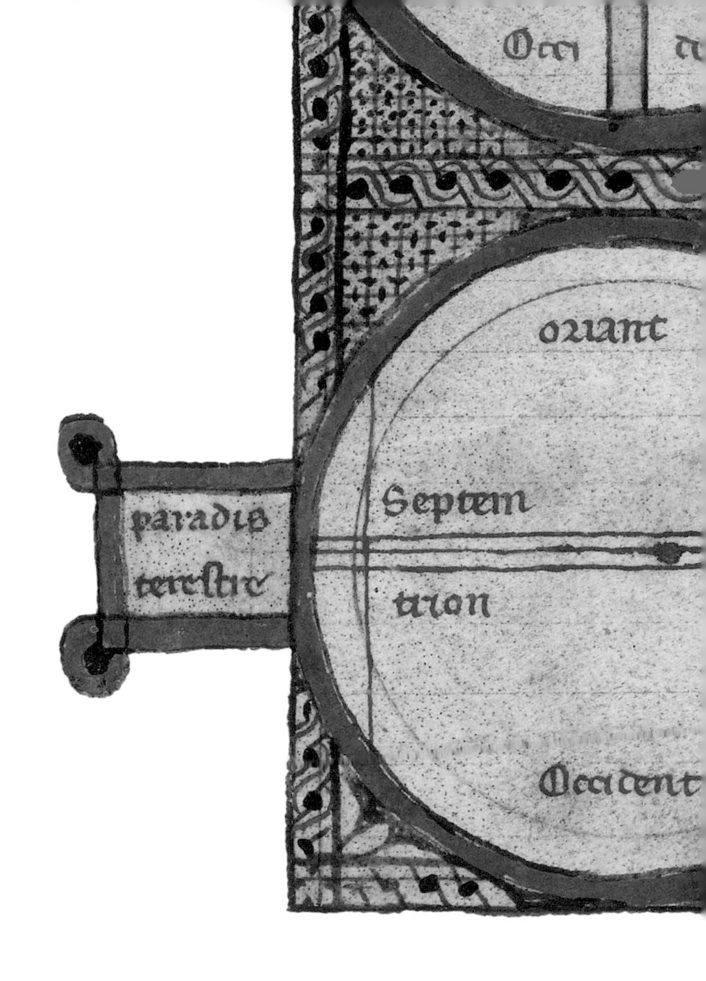

Brunetto Latini, T-O map and map of Paradise, *Le Livre du Trésor*, fourteenth century. MS. Douce 319, fol. iii r

5

A PERSIAN WORLD MAP

The world maps of the late tenth-century Muslim scholar Muhammad al-Istakhrī were copied for centuries, including this beautiful Persian copy dated 1297, painted with gouache and ink on paper. In contrast to Christian medieval maps, it is oriented with south at the top. Africa is shown as a large claw extending deep into the Indian Ocean at top left. The Nile runs from top right downwards, flowing into the Mediterranean with three red circular islands: Cyprus, Crete and Sicily. Unsurprisingly the map's richest detail lies in the Muslim empire, whose various administrative districts are demarcated by red lines, including Spain ('al-Andalus'), the triangle at bottom right.

World map after al-Istakhrī, 1297. MS. Ouseley 373, fols 3b–4a

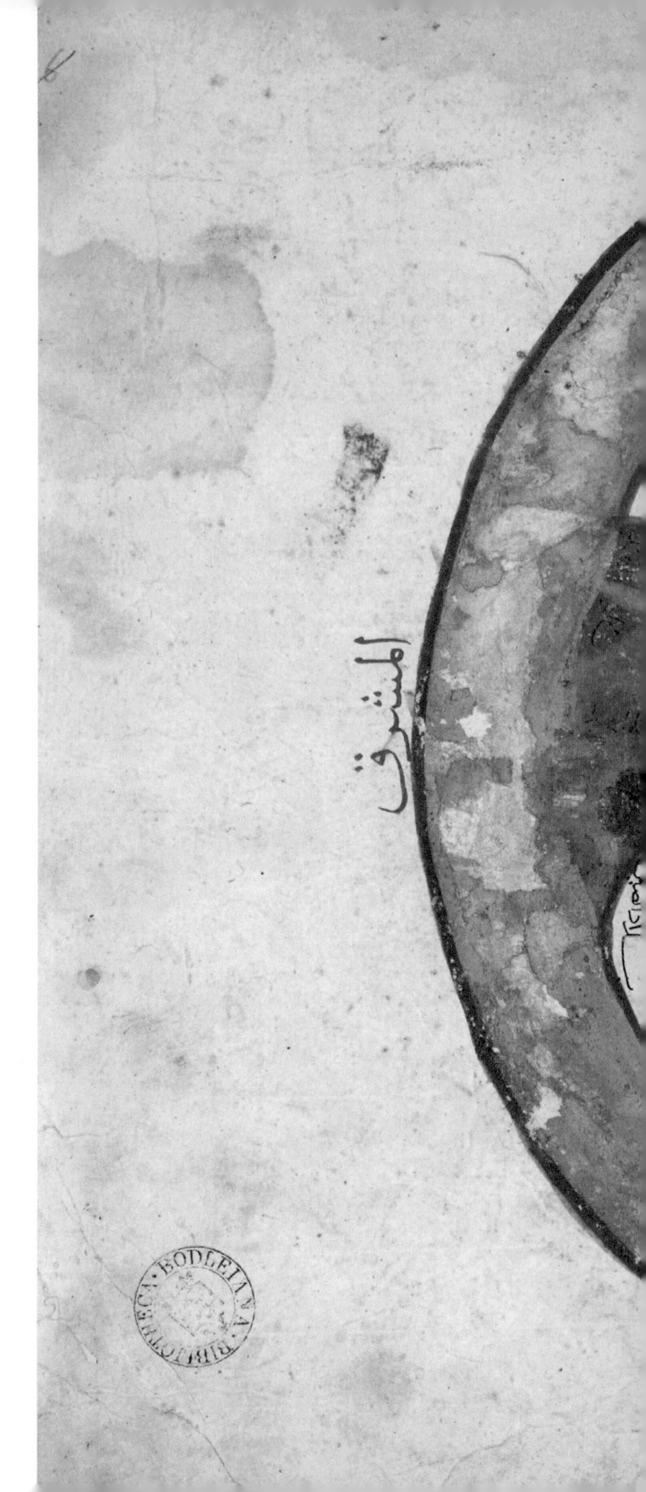

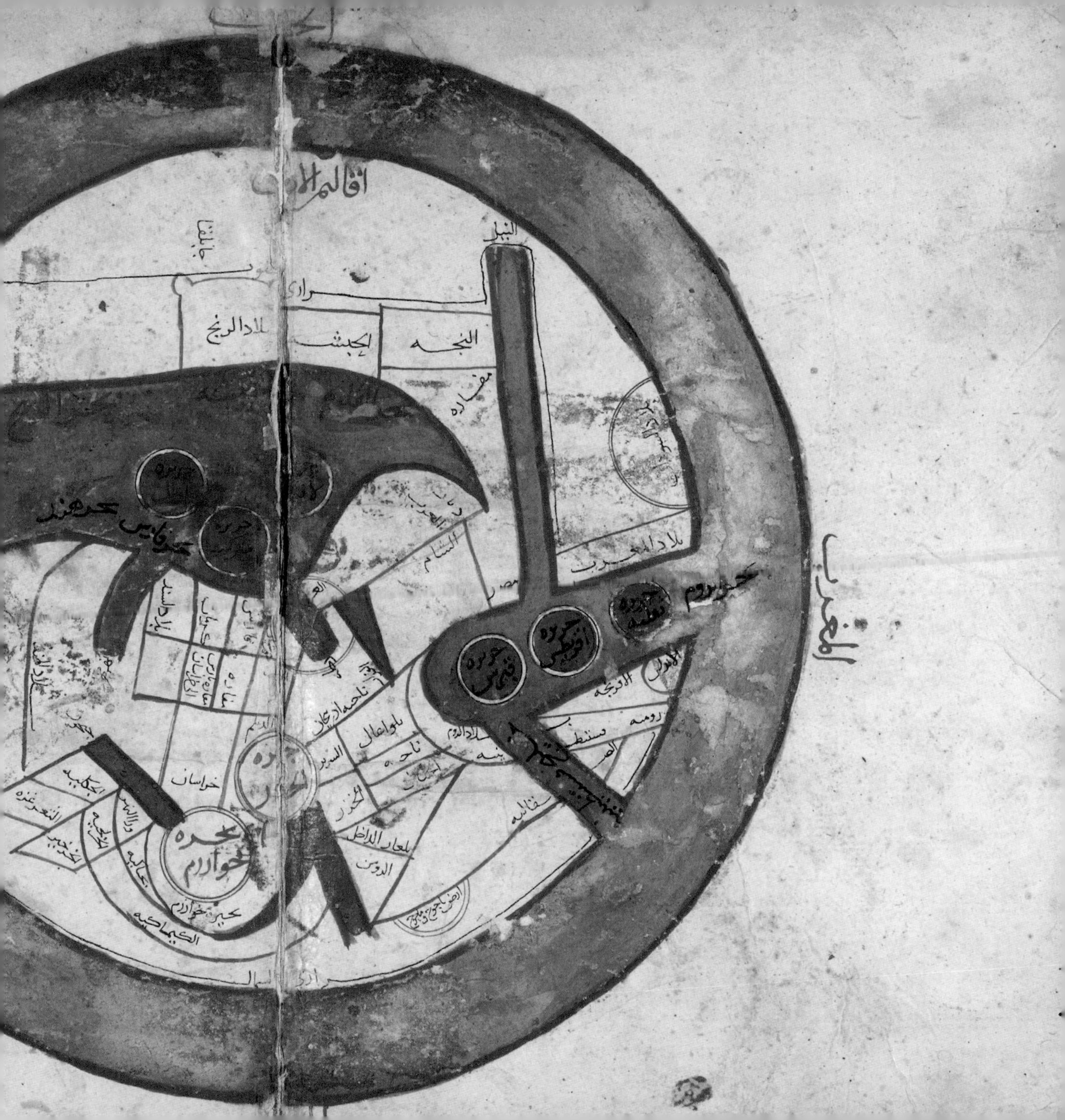
المغرب
اقاليم الارض
النيل
بلاد الزنج
الحبشه
النجه
بلاد المغرب
بحر الروم
جزيره قبرس
جزيره اقريطش
بحر فارس
بحر الهند
الشام
مصر
الاندلس
الافرنجه
رومه
خراسان
الخزر
بحيره خوارزم
خوارزم
الكيماكيه
الروس

gades bachi
hic leones nascuntur
desertum
assiria
cilicia
asya
caucasus
amazones
capadocia
cilicia
flumen eufrate
syria
jerusalem
troya
delos
creta insula
egiptus
gades alexandri
helles pontus
carinthia
alamania
ungaria
europa
grecia
apulia
lombardia
roma
francia
dacia
saxonia
frisia
flandria
toletum
hispania
scotia
anglia
hibernia
mare mediterraneum
gades herculis
africa
cartago
numidia
getulia
mauri

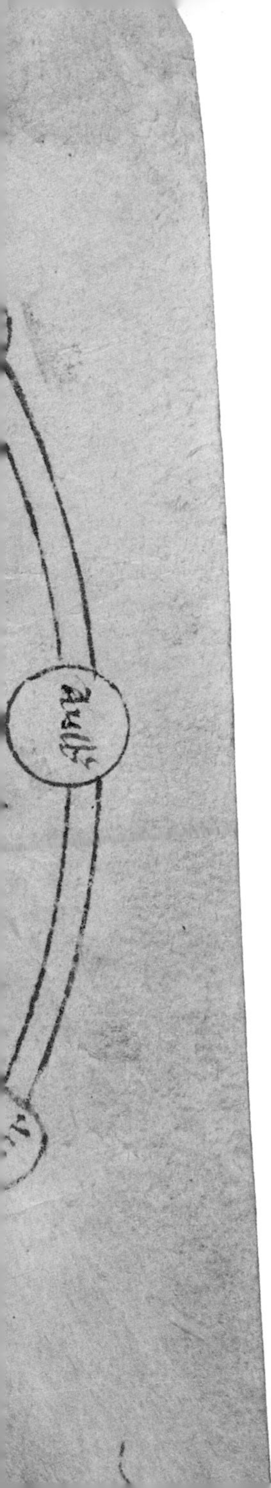

6

A MEDIEVAL T-O MAP

Medieval Christian T-O world maps emerged from a fusion of classical Graeco-Roman geography and biblical cosmology. In them, an ocean encircles the world, with Asia at the top, Europe bottom left and Africa bottom right. Three waterways created the 'T': the Mediterranean running down the lower centre, the Don at the left and the Nile at the right. This exquisite T-O map with a circumference of just 150 mm is from Honorius Augustodunensis' *Imago Mundi* ('Mirror of the World'), a popular medieval encyclopedia of Christian history and cosmology. At its centre is Delos, the birthplace of Apollo in Greek mythology, a symbol of how the map looked to the classical past even as it developed a Christian geography.

Honorius Augustodunensis, *Imago Mundi*, fourteenth century. MS. e Mus. 223, fol. 185r

7

MEDIEVAL *MAPPAE MUNDI*

Mappae mundi (from the Latin for 'cloths' and 'world') offer a remarkable insight into the theology of medieval Christianity. This example was drawn by the Benedictine monk and chronicler Ranulf Higden in his history book the *Polychronicon* (*c.*1350). It is oriented with east and the Garden of Eden at its apex, and Jerusalem, the scene of Christ's crucifixion, at its centre. It is less a geographical route map and more a spiritual guide through the world according to the Bible, starting with Old Testament stories at the top and moving 'down' through the New Testament. This is a world according to Christianity's God.

Ranulf Higden, *mappa mundi* in the *Polychronicon*, fourteenth century. MS. Tanner 170
Opposite Detail of Jerusalem

Babilonia.
Caldea.
Arabia.
Mons
Syna.
Mons
libani
Galilea.
Jerusalem
palestina.
Judea.
Sidon
Idumea
Cedar
Eufrates
Nilus fluvius.
Egyptus
Meroe
Alexandria
Cyprus
Creta
Rodus
Grecia
Italia
colcos
Albania
Gothia
Massagete
asia
minor
Frigia
Galacia
bitinia
pontus
bulgaria
campania
apulia
Rodanus
burgundia
bauaria
saxonia
Westfalia
sicilia
corsica
sardinia
libia cirenensis
mare a
renosum
sirtes
maiores
tripolitana
regio
cartago

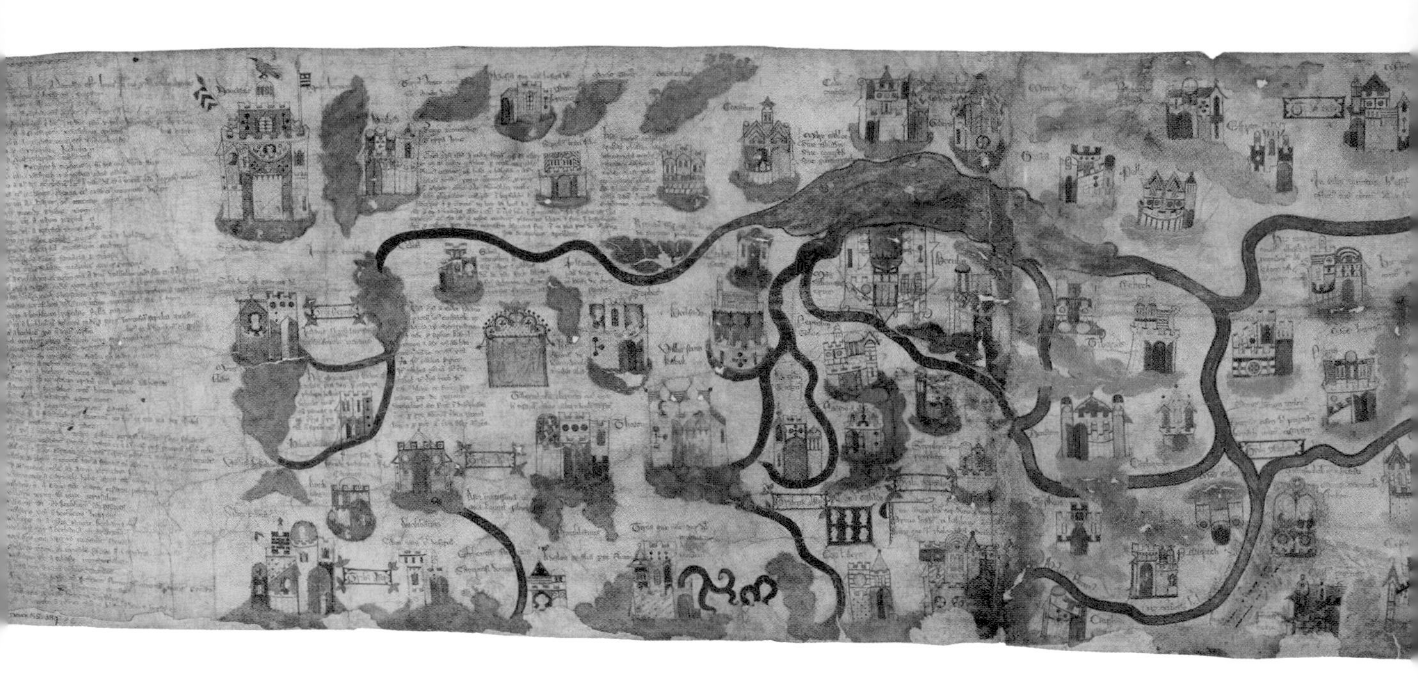

8

PILGRIMAGE TO THE HOLY LAND

Medieval Christian pilgrimage to the Holy Land generated maps to guide the faithful in their journey and to help those unable to travel to spiritually imagine Jerusalem. The Devon priest William Wey described two pilgrimages to the Holy Land undertaken in the 1460s in his *Itineraries* (*c*.1470). Wey seems to have acquired this somewhat earlier map to help him describe his travels. Oriented with east at the top, it runs from Damascus (top left) and Sidon on the Mediterranean coast (bottom left) to the north, to Beersheba and Hebron (far right) in the south. The Jordan River runs across the top, with Jerusalem to the right.

Wey's *Itineraries* match nearly 400 cities, rivers, mountains, sacred sites and even biblical stories found on the map, suggesting he used it as both a topographical and spiritual guide. Next to the castle of Bethel is the legend 'here Jacob saw the ladder'. In the Book of Genesis Jacob comes to Bethel, where he dreams of a ladder connecting the Earth to heaven. The map allowed Wey to both move from one holy site to another, while also seeking a deeper spiritual connection to places where key moments in the biblical story occurred, bringing the past into the present.

Previous page Map of Palestine, late fourteenth century
Opposite Detail of Bethel and Jacob's Ladder.
MS. Douce 389

Jericho

Mons oliuete

Jherusalem

Tribus beniamin

Vall achor

Rama beniamin

Domus zakarie

Modin sepultura machabeorum

9

THE GOUGH MAP OF GREAT BRITAIN

The Gough map is the earliest known example of any map in Britain produced as a separate sheet rather than as a page in a book. It measures 116 x 55cm, is drawn on the hides of a sheep and a lamb, and is unconventionally oriented with east at the top. It is believed to date to the period 1390–1410. Scotland extends out to the left, and is simplistically represented, compared to the more familiar and detailed-looking England and Wales. The map is named after Richard Gough, the antiquary who purchased it in 1776 and bequeathed it to the Bodleian in 1809.

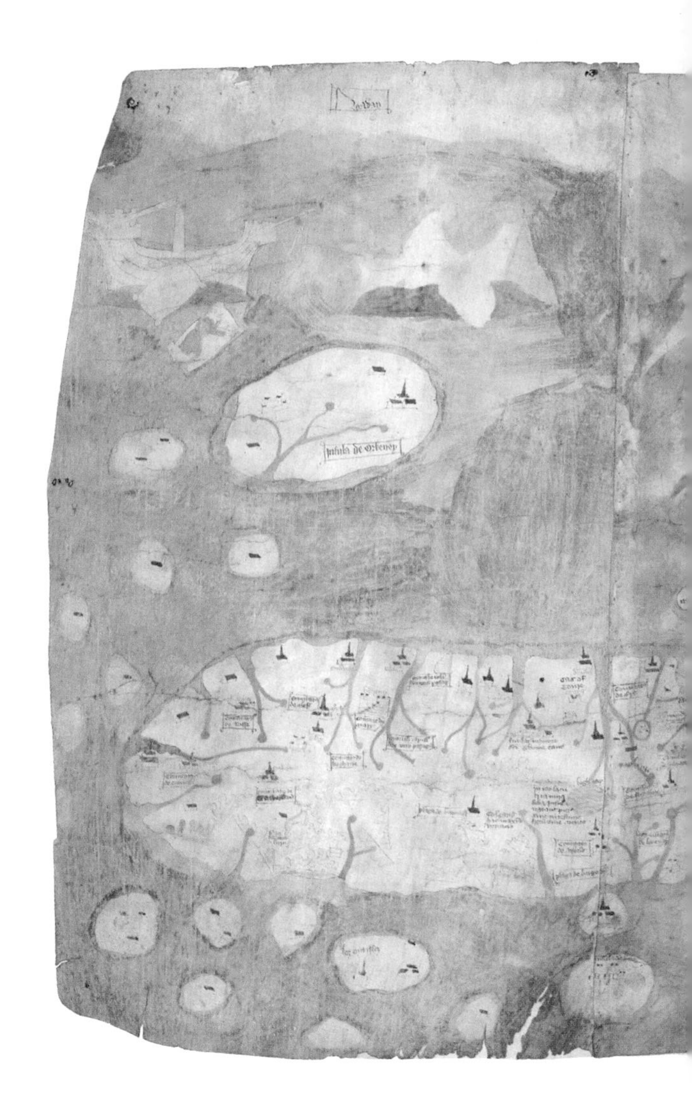

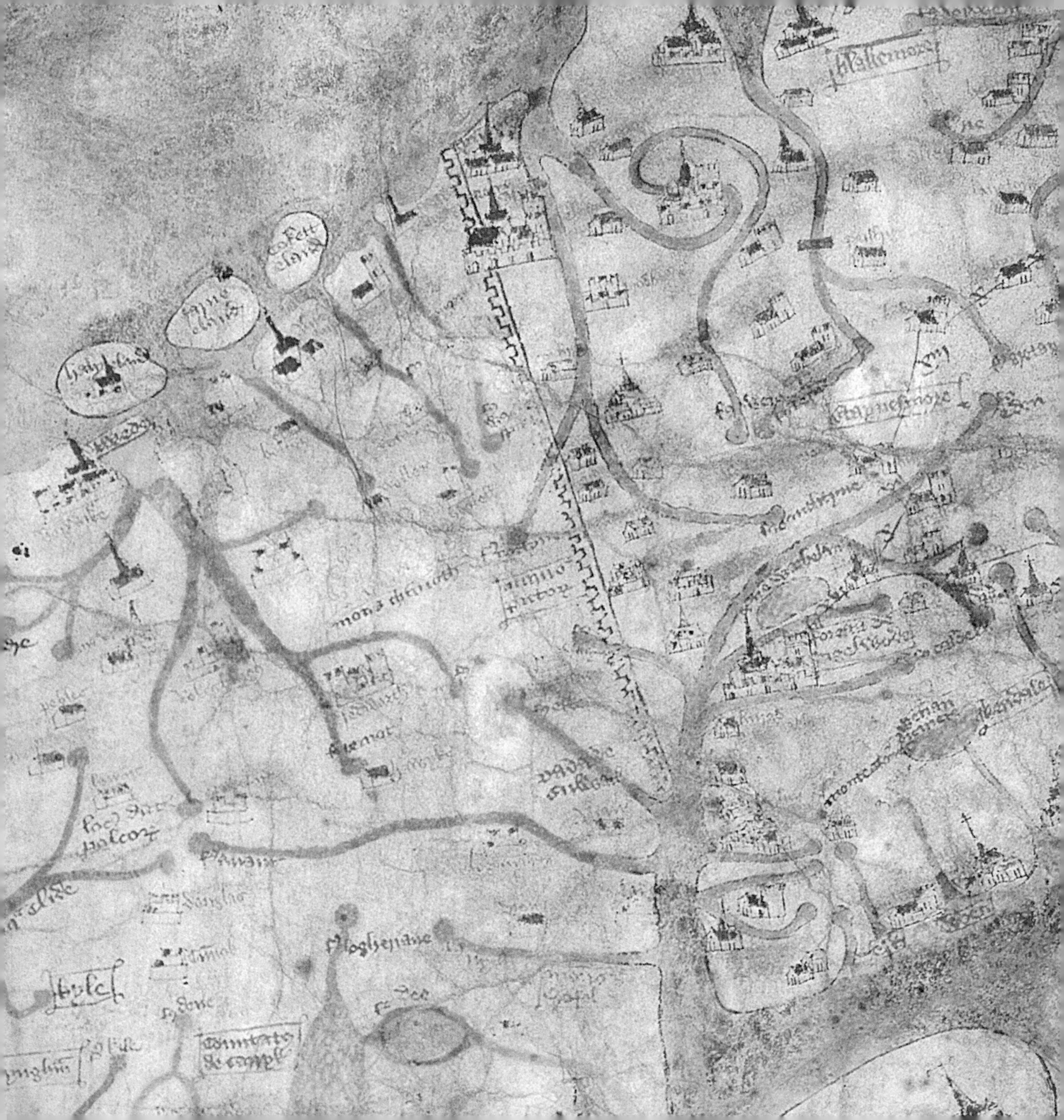

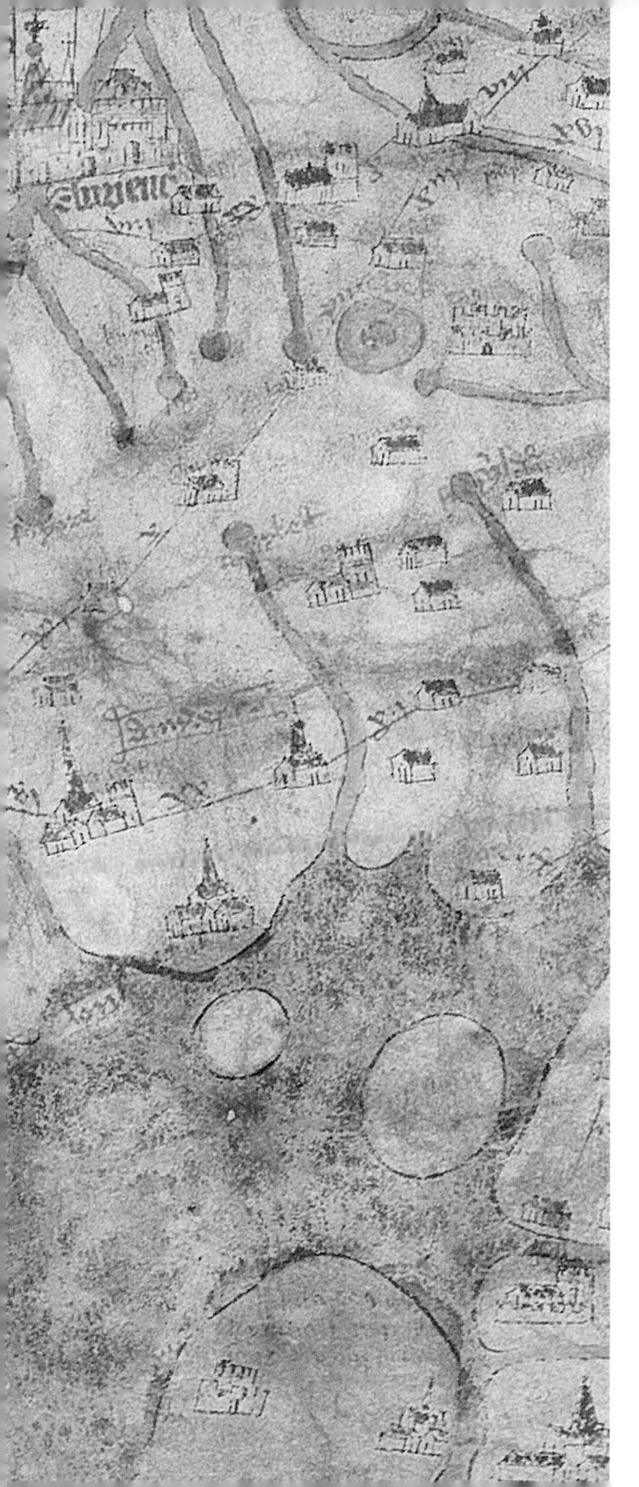

Hadrian's Wall can be seen crossing the North of England in bright red, creating a very real boundary – geographically and, more significantly, practically, in terms of the map's compilation. North of the Wall, in Scotland and most of Northumberland, the scribe's hand is significantly different from the rest of the map, and is older in style. The earliest writing on the map is that found in the red ink text, principally in Scotland, exemplified by the regional names within rectangles. None of the rivers on the northern side have solid darker banks, yet all of those to the south do.

Previous page Map of Great Britain, *c.*1390–1410
Opposite Detail of Hadrian's Wall.
MS. Gough Gen. Top. 16

10

CHARTING ITALY AND THE ADRIATIC

Portolan (nautical) charts of the Mediterranean were an essential possession for any serious medieval merchant or pilot, and many came in richly decorated atlases such as this one, known as the Douce Atlas, containing six charts made in Venice in the fifteenth century. The chart showing Italy and the Adriatic features typical rhumb lines (which cross meridians at the same angle) enabling pilots to navigate by dead reckoning, with place names at right angles to the coast, and no description of the interior. Such atlases focus on maritime and commercial matters, although it is unclear how many of those that have survived were used at sea. Most were probably commissioned to reflect their owners' wealth and social status.

Italy and the Adriatic, portolan chart, fifteenth century. MS Douce 390, fols 5v–6

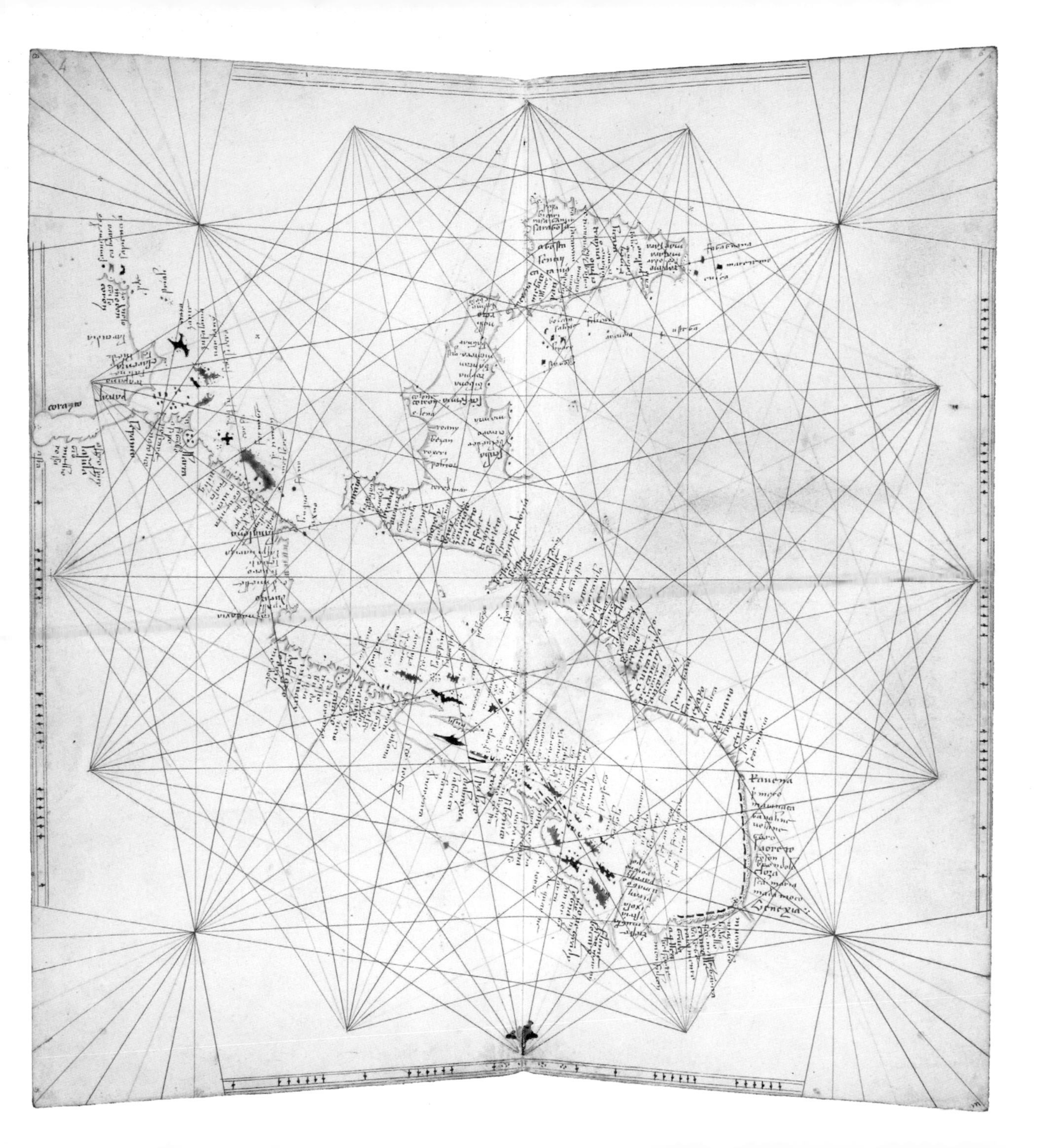

11

DANTE'S HELL

The Italian Renaissance witnessed a fashion for maps calculating the shape and size of hell as described by Dante Alighieri in his poem *The Divine Comedy* (*c.*1308–1320). The Florentine mathematician and architect Antonio Manetti composed this first printed map including Dante's hell, which was published posthumously in 1506. It is shaped like a cone with its *vano* ('void') estimated as 3,245 *miglia* (miles) deep. Hell lies beneath Jerusalem, with Mount Purgatory at its antipodes. Such maps seem to have been designed by Tuscan scholars to celebrate their compatriot Dante and assimilate recent advances in mathematics and geometry with the divine proportions of God's creation – including hell.

Antonio Manetti, *Dante's hell,* in Girolamo Benivieni, *Dialogo di Antonio Manetti* (1506). Toynbee 893

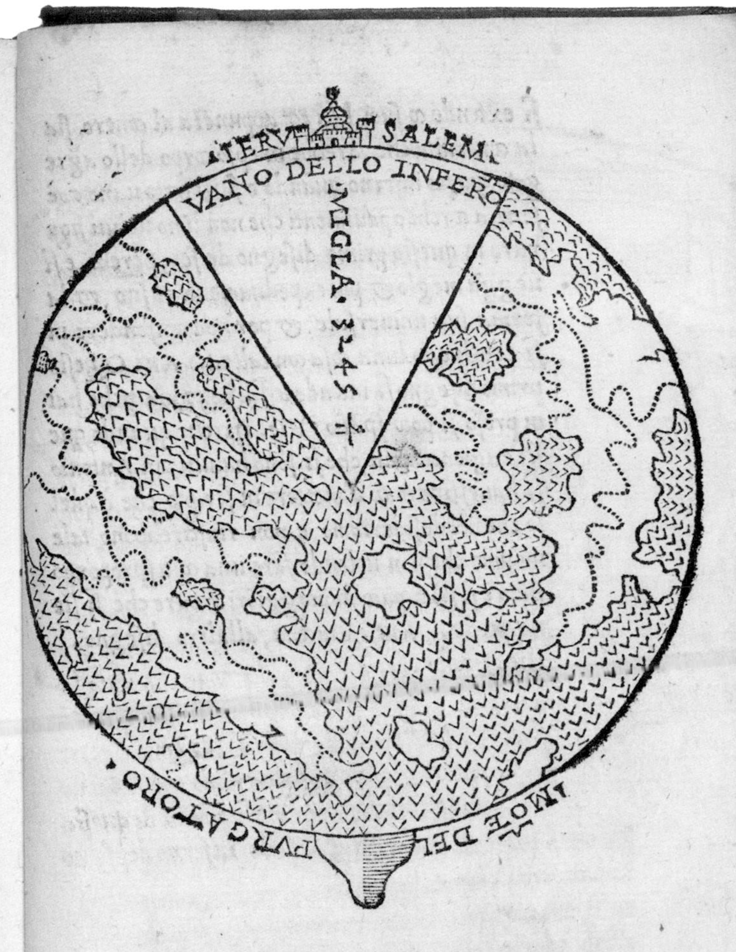

Imaginatiui che questo tondo sia tutto el corpo dello aggregato dellacqua & della terra, & che questo triangolo che occupa (come uoi uedete) la sexta parte di decto aggregato, & che

12

THOMAS MORE'S *UTOPIA*

Thomas More's *Utopia* (1516) describes an ideal commonwealth, but it also puns on the Greek words *outopos*, meaning 'no place' and its homonym *eutopos*, or 'happy place'. Artists such as Ambrosius Holbein were drawn to More's imaginative geography and in this map, from a 1518 edition of More's book, Holbein depicts the island of Utopia with its main characters and places. But Holbein also joins in More's puns and jokes: on closer inspection the island resembles a grinning skull, with the hull of the ship representing teeth and its rigging a nose. Holbein suggests that death and decay await all ideal societies.

Ambrosius Holbein, map of Utopia
in Thomas More, *Utopia* (1518). Wood 639

Amaurotū vrbs.
Fons Anydri.
Ostium anydri

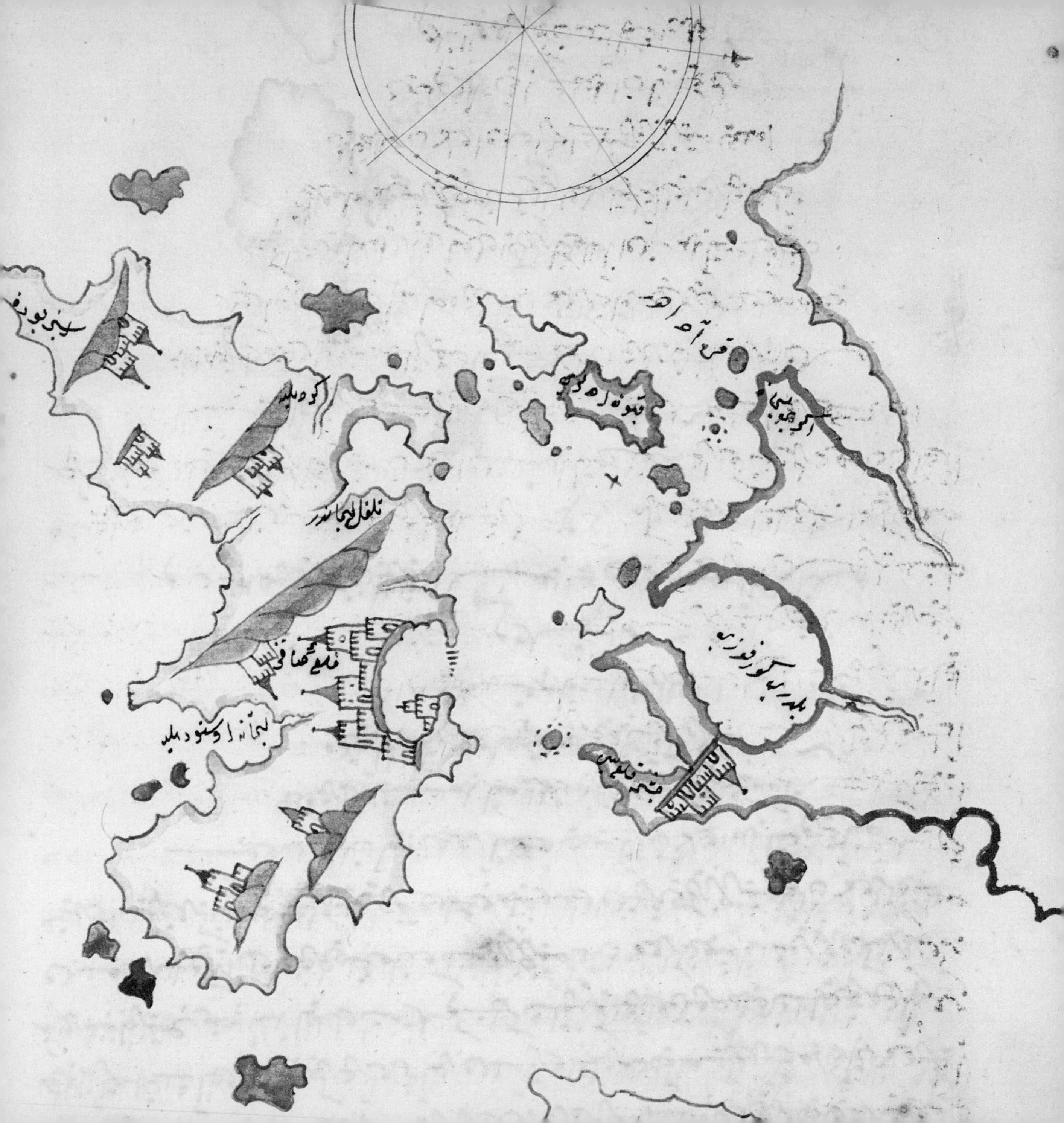

13

THE OTTOMANS MAPPING THE AEGEAN

Sixteenth-century Ottoman imperial sea-power produced some of the period's finest charts of the Mediterranean. These included the naval commander Piri Reis's *Kitab-ı Bahriye* ('Book of Sea Matters') published in 1521 and 1526, a manual describing the region and illustrated with hundreds of exquisite maps such as this one of Chios in the Aegean Sea, just off the western coast of modern-day Turkey. When the map was drawn, the island was a Genoese possession; it fell to the Ottomans in 1566. As well as being a navigational aid, Piri Reis's book was a statement of Ottoman imperial intent in the region.

Piri Reis, 'Saqiz' (Chios), *The Book of Sea Matters*, 1521. MS. D'Orville 543, fol. 19

14

AN AZTEC MAP OF TENOCHTITLAN

In Aztec mythology Huitzilopochtli, the god of warfare and sacrifice, prophesied the foundation of Tenochtitlan (modern-day Mexico City), where his people saw a snake perched on a cactus. This map shows the city in its ideal state, built on lakes and canals and divided into four main zones presided over by its founding fathers, including Tenoch (centre left). Scenes of human sacrifice and battle capture the Aztecs' violent culture. The map was drawn just after the city fell to the Spanish conquistador Hernán Cortés in 1521. As such it represents a world that was about to disappear.

Map of Tenochtitlan, Codex Mendoza, 1542.
MS. Arch. Selden. A. 1, fol. 2r

Acacitli
quapan
ocelopa
aguexotl
tecineuh
tenuch
xomimitl
xocoyol
tenochtitlan
xiuhcaqui
atototl

15

MAPPING MECCA

Islam requires its followers to pray and perform rituals in a sacred direction, or *qibla*, namely facing the Kaaba in Mecca. From the ninth-century CE this led to increasingly elaborate maps based on astronomical and mathematical calculations, including this one of Mecca from a nautical atlas dated 1571 by the Tunisian scholar 'Ali ibn Ahmad Sharafi al-Sifaqsi. It is oriented to the south, with the Kaaba at its centre and a thirty-two-point wind rose radiating outwards to *miḥrābs* (prayer niches) containing the names of Muslim regions or cities. The map could be used as a practical aid to prayer, or imaginatively transport believers to their holiest site.

'Ali ibn Ahmad Sharafi al-Sifaqsi, map of Mecca, 1571. MS. Marsh 294, fol. 4b

ثم ايرة يعرف بها محاريب البلاد وصفة دار استقبال الى الكعبة المشرفة
وقد تقدم صفة استخراج المحاريب وصفة دار استقبال البلاد في الطوان.

قال الله العظيم في محكم كتابه الحكيم وحيث ما كنتم فولوا وجوهكم شطره

16

THE FIRST MODERN COUNTY MAPS

Christopher Saxton, known as 'the father of English cartography', produced the first national atlas of English and Welsh counties, surveying and drawing the maps himself. His work was completely original. Saxton's atlas appeared in 1579, and this Gloucestershire map was completed in 1577. This map measures 48 x 50cm, and was produced at a scale of roughly 1:200,000. We see hills depicted as rounded knolls, standardized symbols for settlements, and bridges providing key river crossings, while forests and gentlemen's seats are also featured. All subsequent county maps for well over a century were based on those made by Saxton (see p. 77).

Christopher Saxton, map of Gloucestershire, *Atlas of the Counties of England and Wales*, 1579. MAP RES 76, fol. 12

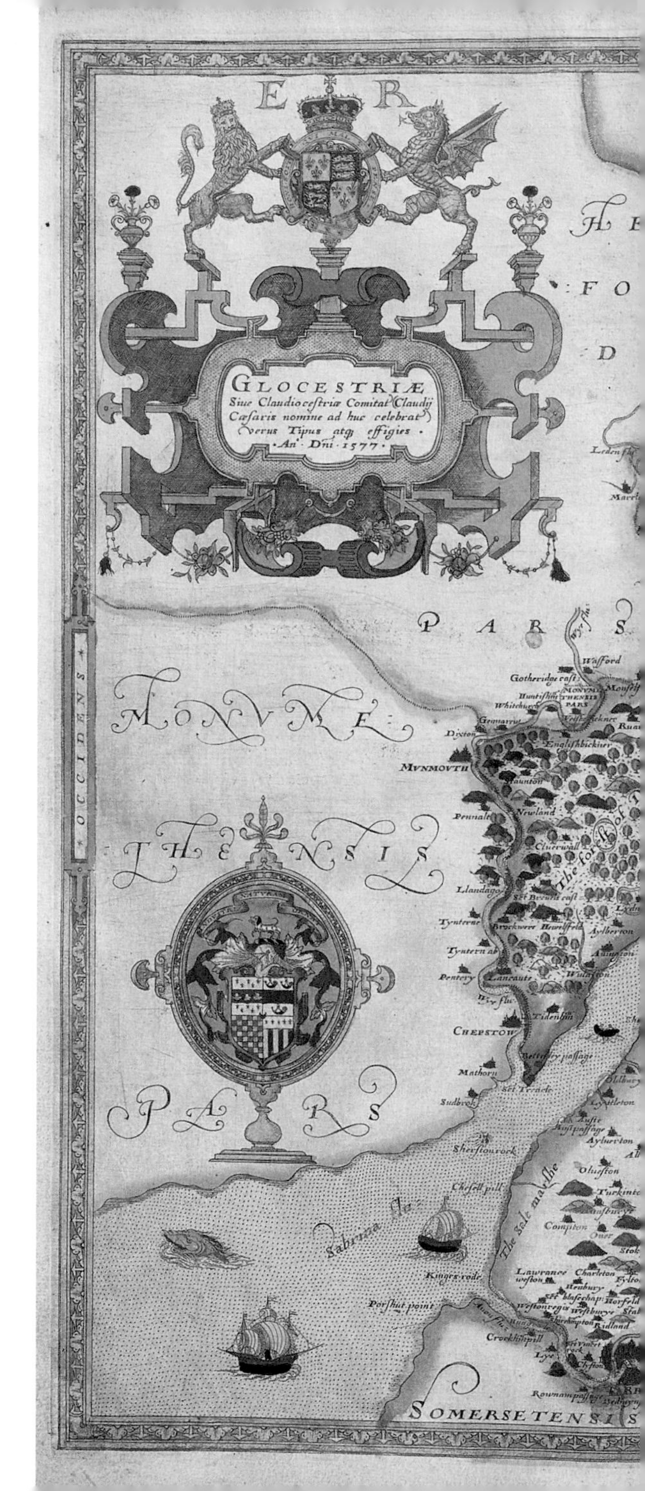

SEPTENTRIO
GORNIÆ
PARS
Sabrina flu:
WARW:
PARS
OXON:
PARS
BERCE:
RIÆ
PARS
WILTONIÆ PARS
ORIENS
MERIDIES
PARS
EVESHOLME
PERSHORE
Auon flu:
THE VALE OF EVESHOLME
CAMPDEN
TEWKESBURYE
WINCHCOMBE
CHELTENHAM
Glocester marshe
GLOCESTER
Cotes Woulde
STOW ON THE WOULD
BURFORD
Euenlode flu:
Windrushe flu:
Stoure flu:
LECHLADE
Thamesis flu:
CIRENCESTER CICITER
CREKELADE
STROUDE
MINCHINHAMPTON
TETBURY
DURSLEY
CHIPPINSODBURYE
MARSFELDE
Scala Miliarium
1 2 3 4 5 6 7 8 9 10 11 12
CHRISTOFERUS SAXTON DESCRIPSIT
AUGUSTINUS RYTHER ANGLUS
SCULPSIT AN DNI 1577

the landes end
mountes baye
Seuen stones
Sillye
The Gulphe
Heilston
Perin
Falmouth hauen
Truro
C O R N
Lisarde pointe
The Spanishe fleete
Den 19en Juny Anno 1588 wert de spaensche vloot
ondeckt van een pynas daer op kappiteyn was
Tomas de Vlaming
Lisarde gewesendt

17

CHARTING THE DEFEAT OF THE SPANISH ARMADA

This is one of a series of ten charts commissioned in the aftermath of the English fleet's defeat of the Spanish Armada in July/August 1588. They provide a vivid account of the various naval engagements over eleven days. The sequence begins with this chart showing the first sighting of the Spanish fleet by an English ship on 29 July; a dotted line indicates the ship's course out towards the massed enemy. The chart's Dutch text describing the encounter suggests it was made in the Low Countries. The victory was so important to the English that several versions of the series were drawn and printed throughout the late sixteenth and seventeenth centuries.

'The Sighting of the Spanish Armada off the Lizard', 29 July 1588. The Astor Armada Drawings, late sixteenth–early seventeenth century

18

TUDOR TAPESTRY MAPS

The Gloucestershire Sheldon tapestry map is woven in wool and silk. This surviving fragment measures 188 x 122.5cm; when complete, the tapestry's original dimensions were around 610 x 460cm. The geographical extent of the fragment is from the northern suburbs of modern Bristol in the southwest, to just beyond Stroud in the northeast. The Forest of Dean, the Severn Estuary and the southern Cotswolds feature prominently. It was originally part of a set of four maps dating from around 1590, heavily based on Saxton's maps (see p. 56), and all commissioned by Ralph Sheldon for his home at Weston, near Long Compton in Warwickshire.

Sheldon tapestry map of Gloucestershire, *c.*1590.
MS. Don. a. 12 (R).
Overleaf Detail of the Forest of Dean

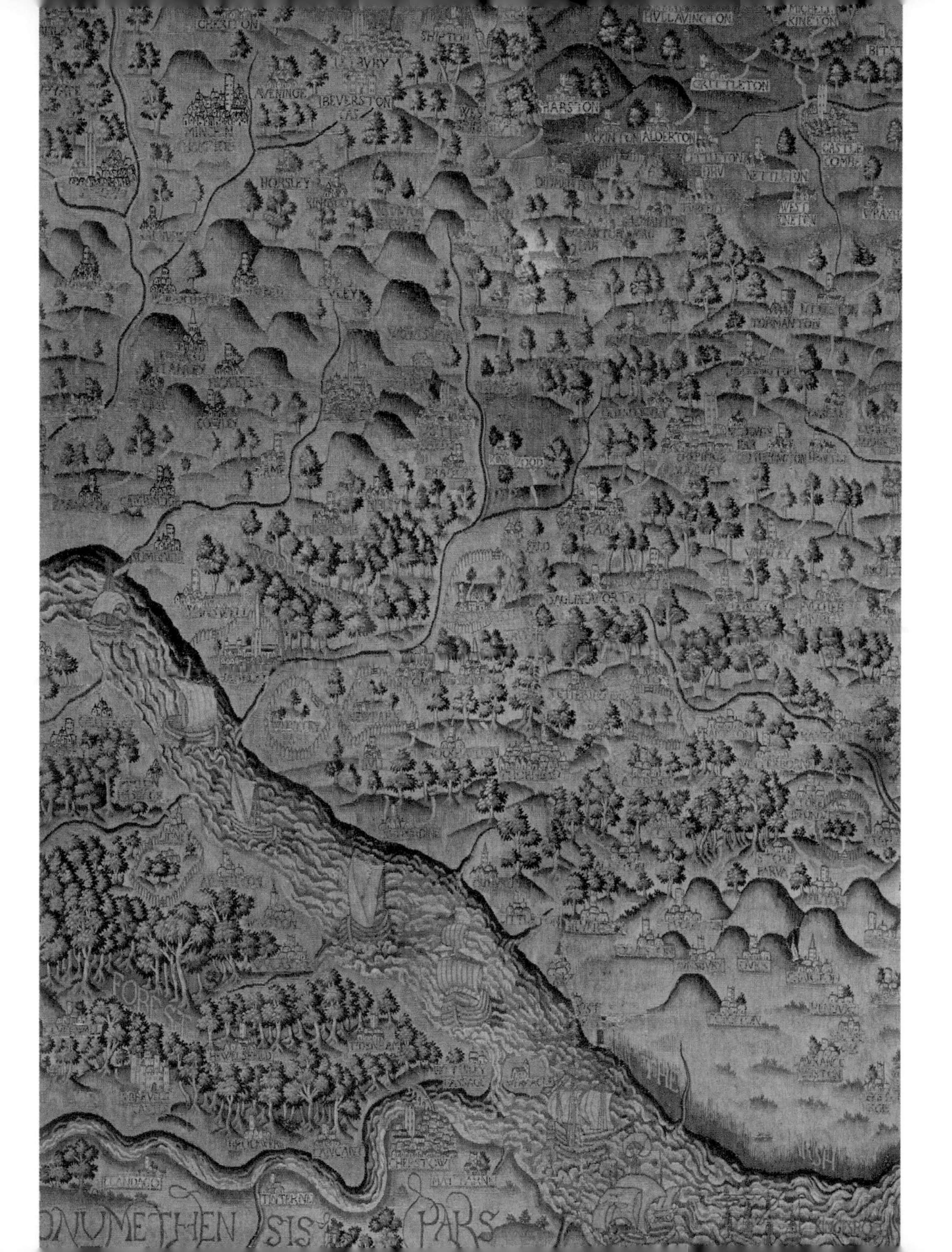

HULLAVINGTON
KINETON
GRITTLETON
AVENINGE
BEVERSTON
HARSTON
ALDERTON
CASTLE COMBE
NETTLETON
WEST KNETON
HORSLEY
KINGSWOOD
TORMANTON
TINTERNE
CHEPSTOWE
MATTHERNE
FOREST
MONUMETHENSIS PARS

HEWELSFELD
PASSAGE
LANCANT
CHESTOW
TINTERNE
HEN
SIS
PARS

On the western side of the map we have the splendid arched crossing of the River Wye (unusual because the tapestry seldom shows bridges); forests are depicted by over-sized trees, soaring above churches and nearby St Briavel's Castle in the Forest of Dean. The Severn ferry can be seen, with two passengers being rowed across to Aust. Each settlement is identified by its own unique vignette, with the number of buildings portrayed being largely in proportion to relative size. Chepstow is one of the grander urban views, observed from the south.

19

AN ELIZABETHAN WORLD MAP

Edward Wright was an English mathematician who published the first mathematical explanation of Gerard Mercator's famous map projection (1569), enabling sailors to use it more effectively. Wright's world map (1599) was defined by its network of rhumb lines, and its assimilation of the latest Spanish, Dutch and English discoveries in the Americas (shown tantalizingly open to further exploration) and Asia. The map's influence even reached Shakespeare, who referred to Malvolio in *Twelfth Night* as smiling 'his face into more lines than is in the new map with the augmentation of the Indies'.

Edward Wright, world map, 1599. (E) B1 (1047)

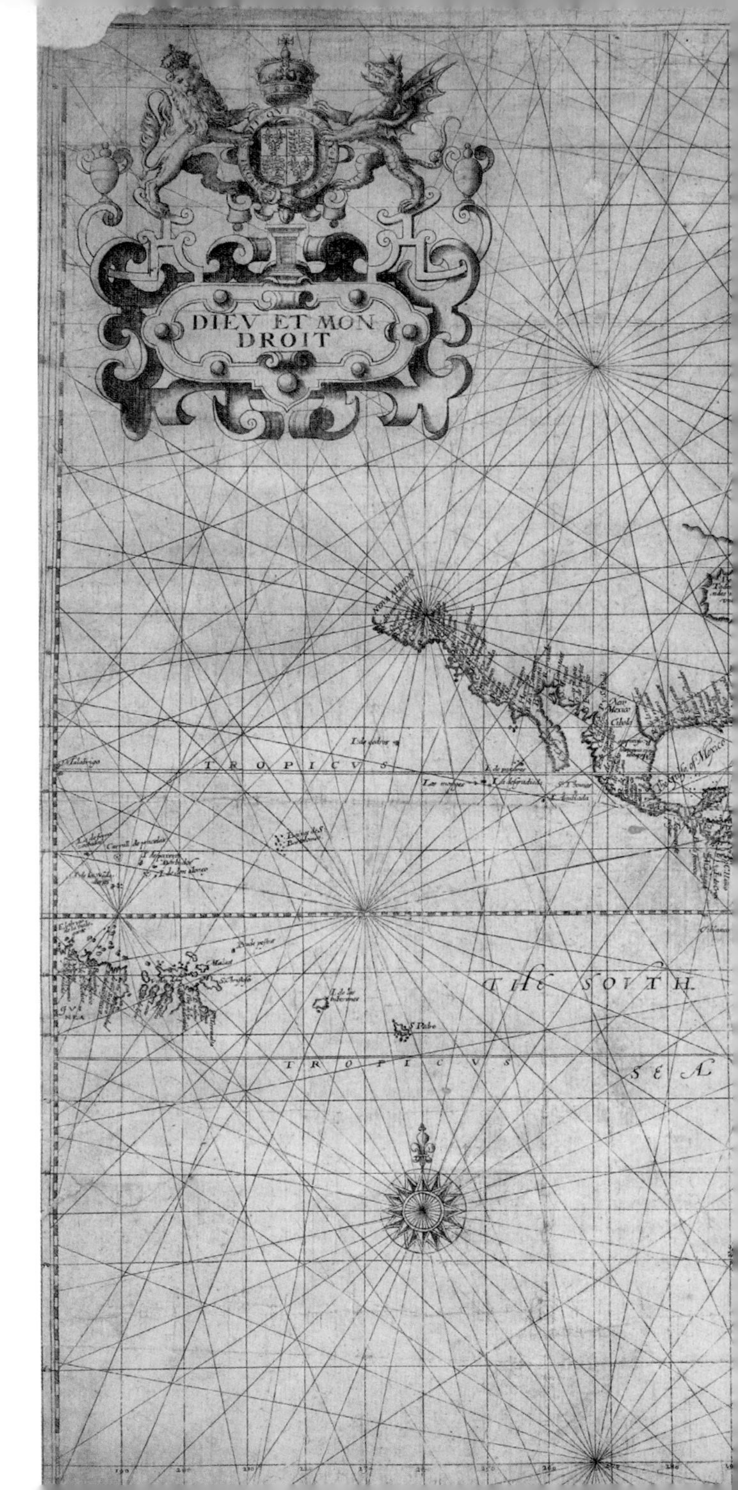

It appeareth by the discouerie of Francis Gaulle
a Spaniard, in the yeare 1584. that the sea betwe-
ene the west part of America and the east of Asia
(which hath bene ordinarily set out as a straight
and named in most maps the streight of Anian) is
aboue 1200 leagues wide at the latitude of 38 degr.
And that the distance betweene cape Mendocino and
cape California which many maps and seacharts
make to be 1200. or 1300 leagues is scarce so much as 600
GRONLAND
Fretum Davis
Islande
NORWAY
EUROPA
Mare Caspium
SAMERCAND
BOCHAR
MOGORES
CHINA
Paquin
TARTARI
CAMBAIA
CANCRI
MAROCO
ARABIA
BRASILIA
CAPRICORNI
Cᵉ bona speranza
NOVA
Thou hast here gentle reader a true hydrographical description of so much of the world as hath
beene hetherto discouered, and is come to our knowledge: which we haue in such sort performed, y
all places herein set downe, haue the same positions and distances that they haue in the globe, being ther-
in placed in same longitudes and latitudes which they haue in this chart; which by the ordinarie sea
chart can in no wise be performed. The way to finde the position, or course from any place to
other herein described, differeth nothing from that which is vsed in the ordinarie sea chart. But to finde the distance; if both places haue the same latitude, see how many degrees of
the meridian taken at that latitude are contayned betweene the two places, for so many score
leagues is the distaunce. If they differ in latitude, see howe many degrees of the meridian
taken about the midst of that difference are conteyned betweene them and so many score
leagues is the distaunce.

20

A REDISCOVERED MAP OF CHINA

Bequeathed to the Bodleian Library in 1654 by the English scholar John Selden, this map of Southeast Asia centred on the South China Sea was neglected for centuries until its recent rediscovery and reassessment as the most significant Chinese map of the last 700 years. Dated to the early seventeenth century, measuring 158 x 96cm and painted with ink on paper by an anonymous Chinese map-maker, it is an exquisite fusion of Ming science, art and commerce. Oriented with north at the top, it shows a maritime trade network reaching from Calicut in India and the Persian Gulf in the west to the spice-producing islands of Indonesia, including Timor in the east (at bottom right).

The Selden map of China, early seventeenth century. MS. Selden supra 105. Detail of India and the port of Calicut (this page)

北京
山東
山西
河南
南京
陝西
湖廣
浙江
江西
福建
四川
貴州
雲南
廣西
廣東
河州
崑崙山
一名崑山
馬湖府
西至黃
河三千里
麗江河西北至
黃河千五百里
正東
正南

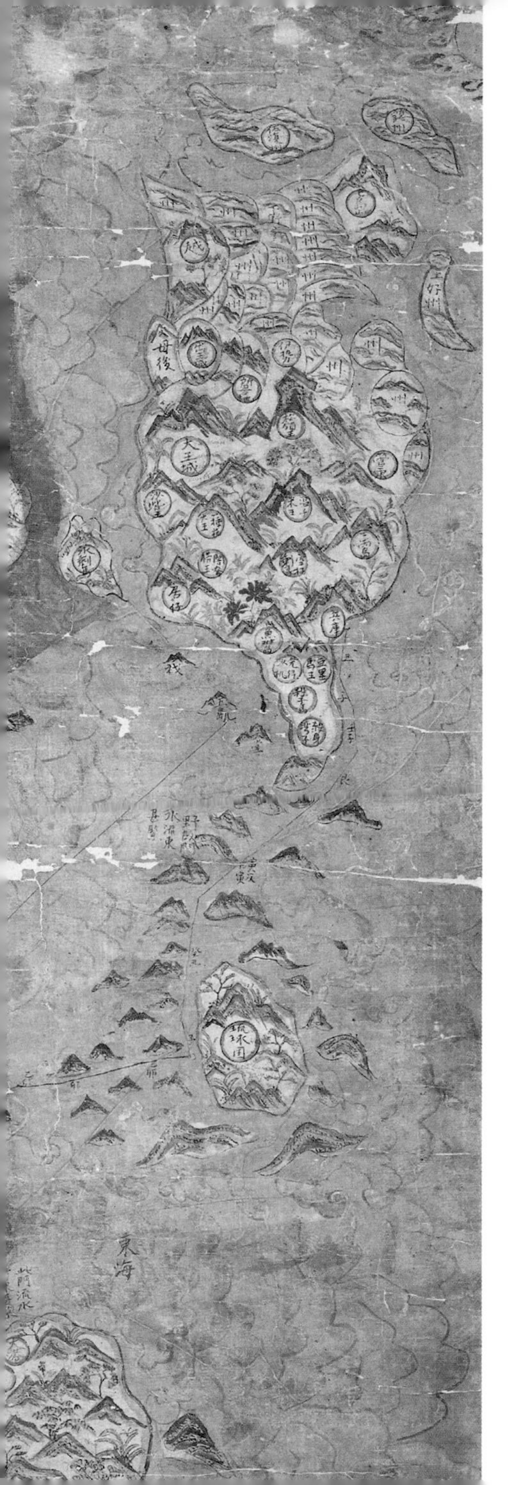

Although the Selden map shows inland topographical features like Beijing and the Great Wall of China, its purpose is maritime trade and navigation. Its faint lines depict more than sixty ports and trade routes that radiate outwards from Quanzhou on the eastern Chinese coast (shown in the centre near Taiwan). The map's coastlines are remarkably accurate, based on measurements using the compass rose and scale bar at the top that represents one Chinese foot (*fen*) at a scale of approximately 1:4,750,000. The rose takes its orientation from compasses (*luojing*) pointing south (*zhinan*), which had been used by the Chinese since the tenth century CE.

Detail of the coastline and the port of Quanzhou showing the compass rose and scale bar on the Selden map of China. Part of the Great Wall of China can also be seen. MS. Selden supra 105

21

FEUDAL RURAL ENGLAND

Measuring 178 x 203.5cm at a scale of 1:3,960, Mark Pierce's manuscript map of 1635 covers the area around Laxton and Kneesall in east Nottinghamshire. The landscape depicted here in such detail is substantially unaltered today. Laxton's pattern of land use – the open-field system – has remained remarkably intact. Three large fields surrounding the village of Laxton survive: Mill Field, South Field and West Field, all discernible on the map (see detail on p. 72). Some 3,333 strips of land, many of them tiny, are marked around the village, each possessing a unique alphanumeric identifier. Although today only a quarter of the current parish remains as open fields, Laxton's land management remains unique in Britain.

In the second detail from the Laxton map (see p. 73) we see people busy working in the landscape, engaging in harrowing, harvesting, hay-making, milking, mowing, ploughing, scything, shepherding and sowing. These activities do not all take place at the same time of year, so the scenes need to be viewed with care. We also witness people indulging in rural sporting activity, hunting deer and hares, as well as hawking. Aside from those animals being hunted, domesticated creatures are visible in the fields – cattle, horses and sheep. Here we also see ploughing in Kneesall's Mill Field.

Laxton map by Mark Pierce, 1635. MS. C17:48 (9)

KIRTON
LORDSHIP
OF
N
LO
OSSINGT
Y
ABB

SOVTH
FIELD
MAYNWOOD

KNEESALL
MILL
FIELD

22

'FALSE' MAP OF OXFORD

Made in the 1640s during the English Civil War, this map shows Oxford at the top of the page; below, the map's scale is drastically reduced, the city tethered by the River Thames to 'Abbington' (Abingdon), Wallingford and 'Reding' (Reading). At the foot of the page lies a manuscript addition by the late seventeenth-century Oxford antiquary Anthony Wood, stating: 'This map is made very false'. Close inspection reveals why: the urban layout is a mirror image along its north–south axis, the street pattern having been taken from John Speed's 1611 map and then 'reversed', so St Giles lies to the south, St Aldate's to the north. Might this map have been created to confuse besieging enemy forces?

Civil War map of Oxford, 1644. Wood 276b, fol. XXX

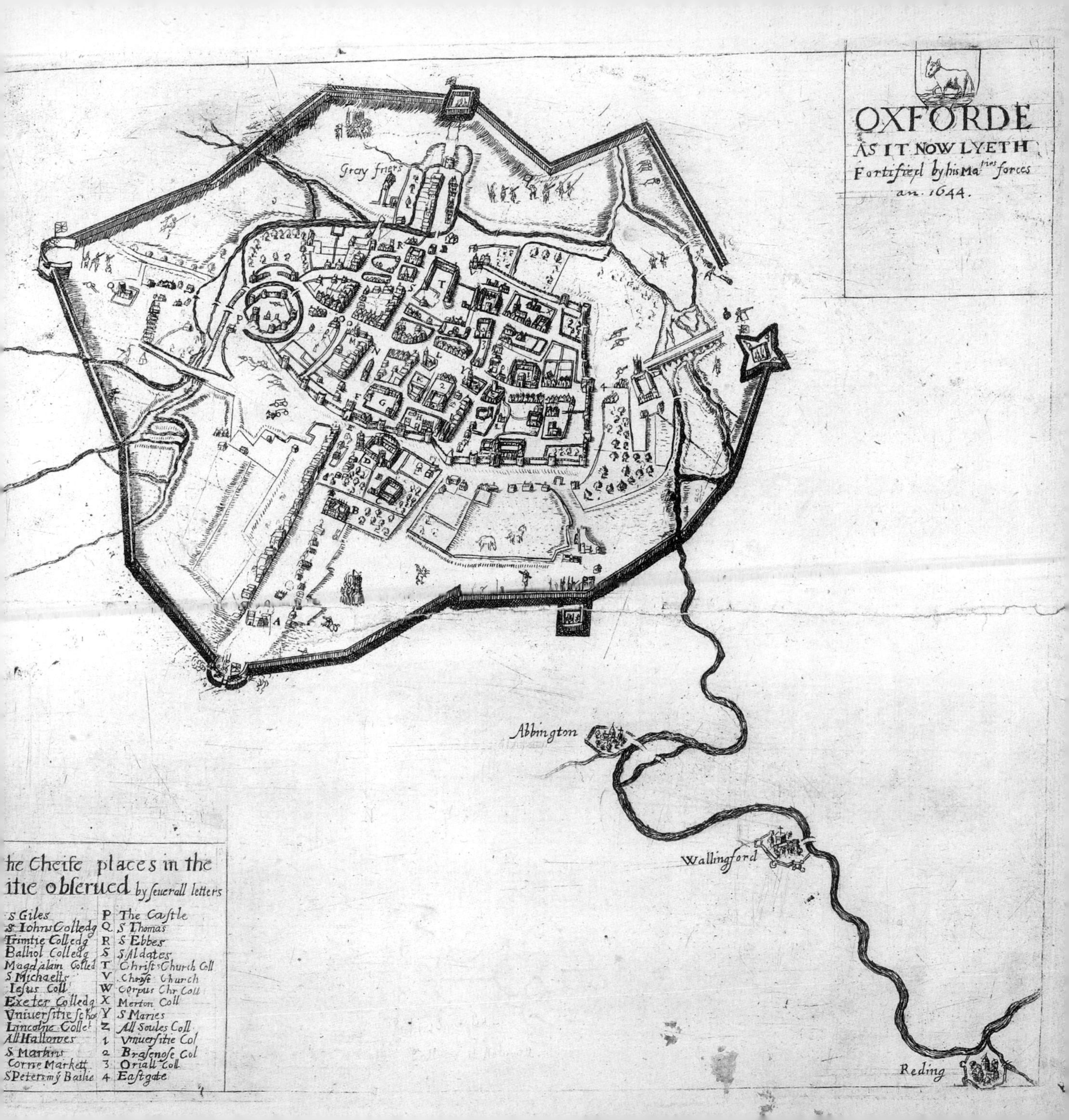
OXFORDE
AS IT NOW LYETH
Fortified by his Maties forces
an. 1644.
Gray friers
Abbington
Wallingford
Reding
he Cheife places in the
itie obserued by seuerall letters
S Giles
S Iohns Colledg
Trinitie Colledg
Balliol Colledg
Magdalain Colled
S Michaells
Iesus Coll
Exeter Colledg
Vniuersitie Schoo
Lincolne Colle
All Hallowes
S Martins
Corne Markett
S Peters in ye Bailie
P The Castle
Q S Thomas
R S Ebbes
S S Aldates
T Christs Church Coll
V Christ Church
W Corpus Chr Coll
X Merton Coll
Y S Maries
Z All Soules Coll
1 Vniuersitie Col
2 Brasenose Col
3 Oriall Coll
4 Eastgate

Septentrio
WIGORNIENSIS
COMITATVS PARS;
PART OF WORCHESTER SHIRE.
HEREFORDIÆ PARS;
HEREFORD COMITATVS
HEREFORD SHIRE.
Occidens
MONVMETHENSIS COMITATVS PARS;
MONMOUTH SHIRE.
Rob. Fitz Hamon Erle of Glos.
Robert D'Aillene E. of Gl.
Richard Mclare E. of Glo.
Hugh Spencer Erle of Glo.
Hugh D'Audley E. of Glo.
Humfry Duke of Glo.
Richard Duke of Glo.
Malvern Hills
Pershore
Tewkesbury
Tewkes byrye hvnd.
Cleve hvnd.
Winchcomb
Cheltenham
Cheltenham hvnd.
Lidbyry
Newent
Botlow hvnd.
Dean mag.
The Forest of Dean
S. Brevells hvnd.
Blideslow hvnd.
Newnham
Westbury
Dvdston
Glocester
Kings Barton hvnd.
Rapis gate hvnd.
Whitston hvnd.
Besley hvnd.
Stroud
Minchin hampton
Longtre
Tetbury
Crothorne
Cirencester
Barkley
Barkley hvnd.
Dursley
Wotton hvnd.
Micklewood Chase
Wotton Underedge
Thornbvry
Thornbury hvnd.
Henbvry
Svynshead & Langley hvnd.
Chippingsodbury
Puckle chvrch hvnd.
Pucklechurch
Barton hvnd.
The Forest of Kingeswood
Bristol
Bidminster
Canesham
Monmouth
Dixton
Whitchurch
Staunton
Newland
Llandogo
Tyntern
Tyntern abb
Pentery
Wye fl.
Chepstow
Mathorn
Sudbrok
Sherston Rock
Chesel Pill
Kinges rode
Porshut poynt
Sabrina Flu.
Severne Fluvius
The salt Marshe
Avon fl.
Crockhampil
S. Vincents rock
Clifton
Rownam passage
SOMERSET SHIRE.
Meridies
WI
PART

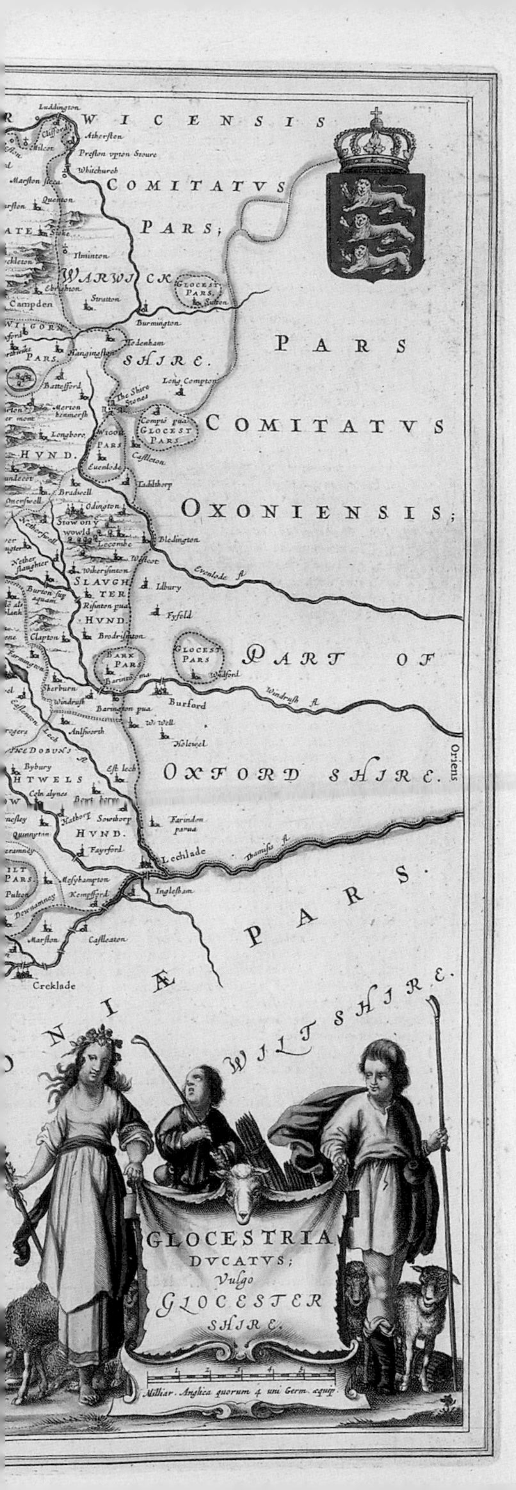

23

BLAEU'S MAP OF GLOUCESTERSHIRE

Joan Blaeu's map of Gloucestershire bears distinct similarities to Saxton's (see p. 56). It appeared in his atlas of England, and the illustration here is taken from the French edition, the fourth volume of *Le Théâtre du Monde ou Nouvel Atlas*. The map measures 40 x 50cm. Being produced at a slightly larger scale than Saxton's map renders it less cluttered, given that Blaeu altered little of Saxton's detail – his hills are rather less rounded and his trees appear more realistic – although he chose to add administrative boundaries. Also featured is a splendid cartouche adorned with sheep, emphasizing the prominence of wool in the Gloucestershire economy.

Joan Blaeu, 'Glocestria Ducatus; Vulgo Glocestershire', from *Le Théâtre du Monde ou Nouvel Atlas*, 1645. MAP RES 31

24

RESTORATION OXFORD

David Loggan's map of Oxford is a *tour de force* exemplified by his interpretation of the city's buildings. It was produced at a large scale of around 1:3,250 (almost twenty inches to the mile). Aside from the map's orientation with south at the top, Loggan also chose to emulate the bird's-eye perspective employed by land surveyor Ralph Agas for his own plan of Oxford in 1578. Unlike traditional maps, Oxford's buildings are given a three-dimensional appearance, with their relative heights shown as well as their locations on the ground. The combination of urban dwellings, grand university buildings and ornamental gardens gives a vivid sense of what 1670s Oxford looked like.

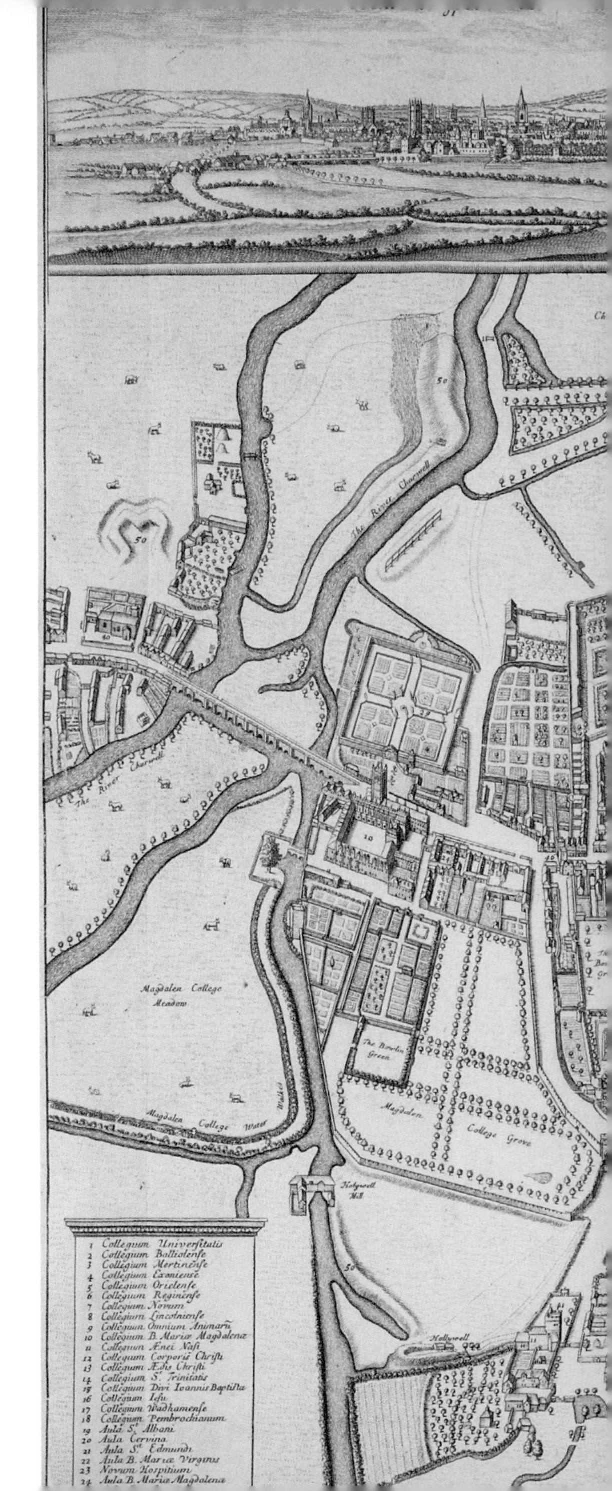

David Loggan, *Nova & accuratissima celeberrimae universitatis civitatisque Oxoniensis scenographia*, 1675. (E) C17:70 Oxford (12)

Reverendissimo in Christo
Patri, natalium splendore, vir-
tutum meritis, literarum Scien-
tiâ, Sacris demum infulis consumma-
tissime Illustri, D.no HENRICO COMPTON
Episcopo Oxoniensi; sedes suæ (quæ
tanto Præsule quasi novo fastigio
aucta altius assurgit) Ichnogra-
phiam hanc in obsequij debitissi-
mi tesseram D. D. C. Q.
Dav. Loggan.
The Black Fryers
The Castle
Broken Hays
Beaumont
The white Fryers
Glocester green
Rusty House
1 University College
2 Baliol College
3 Merton College
4 Exeter College
5 Oriell College
6 Queens College
7 New College
8 Lincoln College
9 Allsoules College
10 Magdalen College
11 Brazen-nose College
12 Corpus Christi College
13 Christ Church College
14 Trinity College
15 S.t Iohns College
16 Iesus College
17 Wadham College
18 Pembrock College
19 Alban Hall
20 Hart Hall
21 Edmund Hall
22 S.t Mary Hall
23 New Inn
27 The Publick Library
28 The Theater
29 The Physick Garden
30 Christ Church Almshous
31 S.t Maries Church
32 Carfax
33 Allhollowes
34 S.t Aldats
35 S.t Ebbs
36 S.t Peters in the Bayly
37 S.t Michaels
38 S.t Magdalen
39 S.t Peters in the East
40 S.t Clements
41 Hollywell
42 S.t Giles
43 S.t Thomas
44 The Town Hall
45 Bocardo and North gate
46 The East gate
47 Frier Bacons Study
48 Paradise garden
49 The Gray Friers

67

༄༅། །ཏེ་ལམ་

སྤྱི་རིགས་ཀྱི་བུ་༧ ལྷོ་ནས་

ཤང་མགོ་ལགན་སྐྱུ་གྲ་ཕྱོགས

སྤྱིའི་མགོ་ལགན་བུམ་པ་ཕྱོགས་པ

པའི་མགོ་ལགན་པདམ་ཕྱོགས་

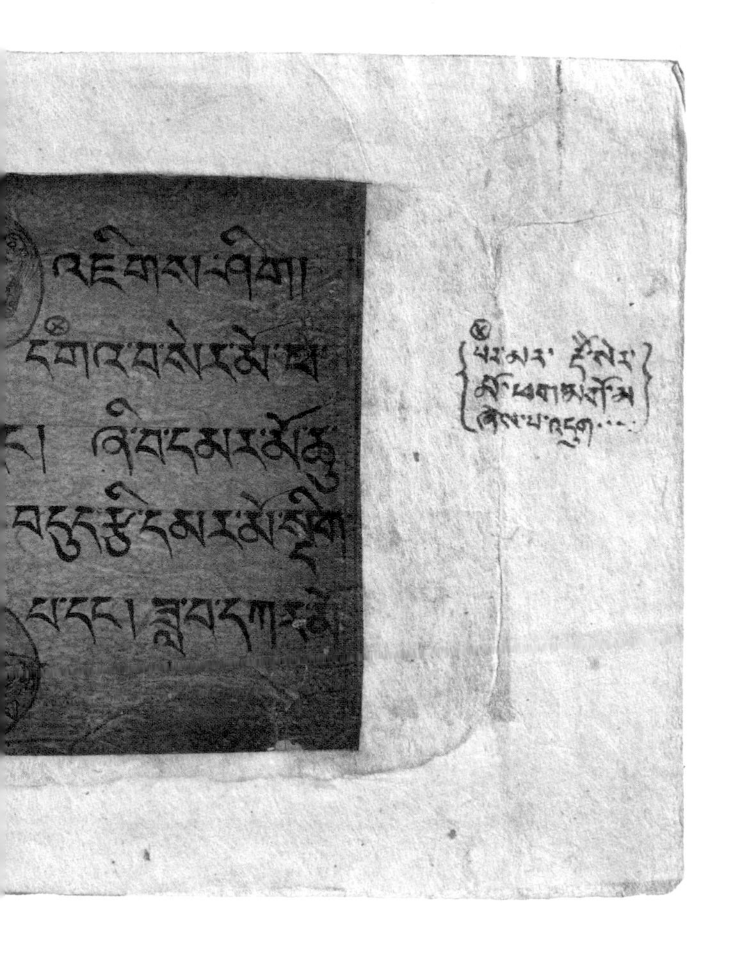

25

THE TIBETAN BOOK OF THE DEAD

One of the most influential yet enigmatic texts popularizing Tibetan Buddhist accounts of the afterlife is the *Bardo Thödrol*, or 'The Great Liberation through Hearing in the Intermediate State'. Translated into English as the *Tibetan Book of the Dead* (1927), it is a funerary manual shown to the dying to help the soul navigate through the liminal world (*bar do*) between death and rebirth. As shown here, it prepares the dying for the hallucinogenic visions and extreme emotions they must endure on their journey. It offers a spiritual route map to salvation, on very different terms from those of Christianity.

Tibetan Book of the Dead, eighteenth/nineteenth century. MS. Tibet c. 60 (R), fol. 67

Aust
Elberton
Olveston
Tockington
Alveston
Tytherington
Cromhall
Northwick
Redwick
Almondsbury
Iron Acton
Frampton Cotterell
Yate
Chipping Sodbury
Winterbourn
Stoke Gifford
Filton
Westerleigh
Codrington
Compton Greenfield
Henbury
Westbury
Horfield
Stapleton
Mangotsfield
Pucklechurch
Siston
Abstone
Doynton
Kingswood
Redland
Durdham Down
Leigh Wood
Abbots Leigh
Leigh Court
Ashton Court
Long Ashton
Bedminster
Brislington
Bitton
Londonderry
River Avon
Sold by Jas. Gardner Agent for the Sale of the Ordnance Maps 163 Regent Street London.
Scale of Statute Miles.
Furlongs
Miles
Engraved at the Ordna

26

THE FIRST ORDNANCE SURVEY MAP OF SOUTH GLOUCESTERSHIRE

A truly national map was conceived following Ordnance Survey's foundation in 1791. The project to map the whole of England and Wales to standard specifications at a scale of one inch to one mile saw its first map, covering Kent, published in 1801. Gradually these maps extended northwards, and became known as the 'Old Series'. These maps performed the same general function as a modern Landranger, the aim of both series being to map the whole country at an identical scale. Measuring 78 x 62cm, sheet 35 (in its entirety) covers the part of south Gloucestershire almost identical in geographical extent to the Sheldon tapestry map (see pp. 60–63).

Detail of Ordnance Survey Old Series (first edition) one-inch maps of England and Wales, sheet 35, 1830. Allen 385

हरिसलिलानदी
निषधपर्वत
गंधापाती
हरिकांतानदी
रोहितानदी
महाहिमवंत
पर्वत
निषधहरिवर्षक्षेत्र
महाहिमवंत
रोहितांशानदी
रूपकूट
गंगानदी
विलंधर

27

A JAIN MAP OF THE UNIVERSE

Jainism is one of India's oldest and most complex religions. Its belief in the soul's rebirth (or transmigration) within an eternal and multi-dimensional universe (*Loka*) requires intricate and precise cosmological maps such as this one to enable its followers to navigate physically and spiritually. It depicts the terrestrial world (*Jambudvipa*) centred on the holy mountain of Meru (possibly the central Asian Pamir mountains). Humanity dwells in the sixteen rectangular provinces (*videhas*), beyond which lie spiritual 'realms' where creatures metamorphose and transmigrate, guided by omniscient human teachers (*Jinas*). Such maps were displayed in temples as spiritual guides for the faithful.

Jain map, sixteenth–seventeenth century.
MS. Or. Evans-Wentz 1

28

SIBERIAN SEALSKIN MAP

Variously described as a drawing, painting or pictogram, this intriguing object can also be called a map. Dating to around 1860, it was drawn on sealskin by a member of the Chukchi people, semi-nomadic inhabitants of the Asian side of the Bering Strait in Siberia. Chukchi life was built around hunting, fishing and trade in furs and skins. All are depicted here, from the slaughter of whales and reindeer to encounters with European whaling ships (bottom right). It is a geographical map of Chukchi territory – showing settlements at St Lawrence Bay (bottom left) and Plover Bay (bottom right) – as well as a record of their everyday life and customs.

Siberian sealskin map, *c.*1860,
Pitt Rivers Museum 1966.19.1

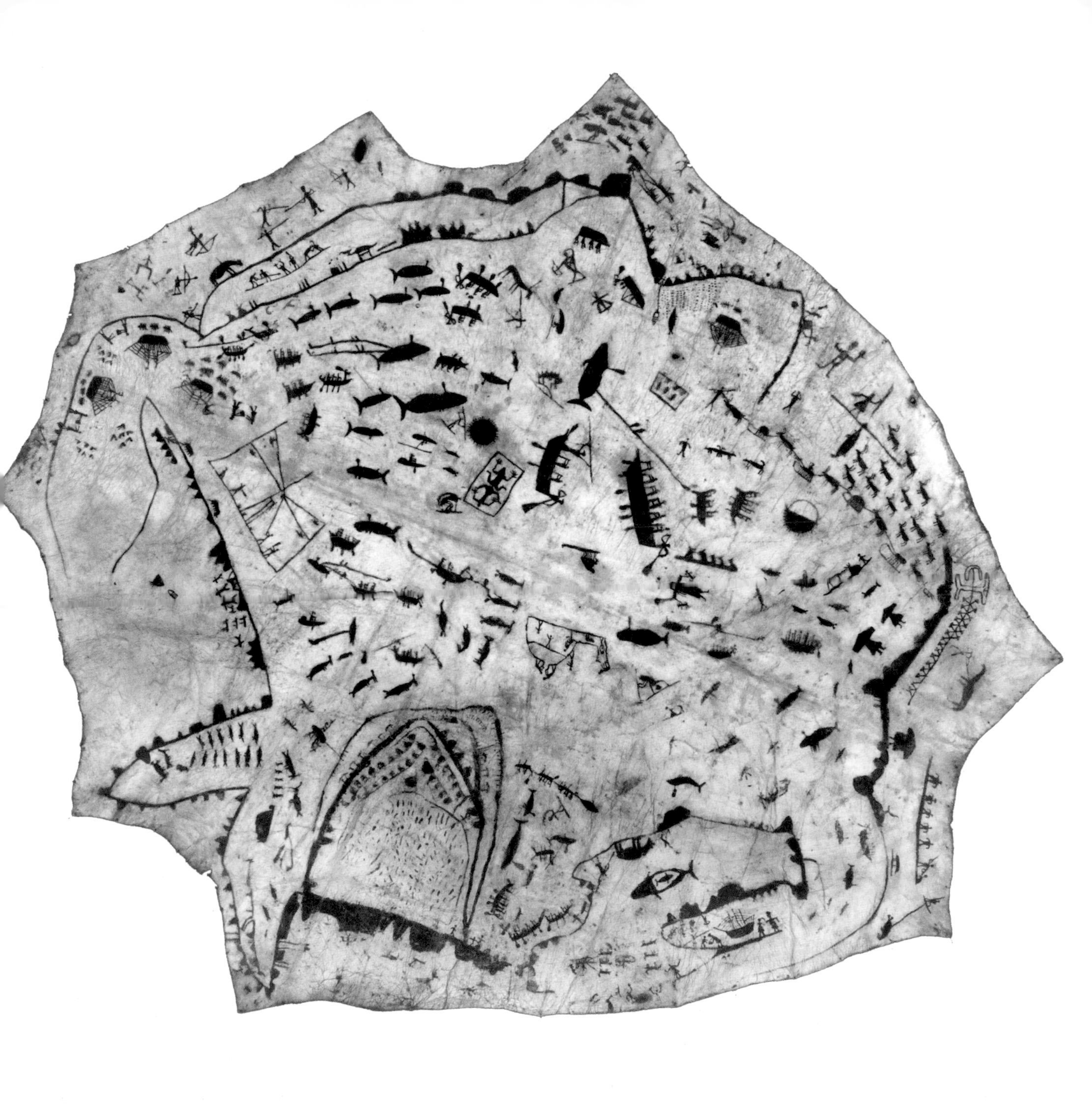

29

FRANCO-PRUSSIAN WAR COMIC MAP

Occasionally, military conflict can act as an artistic catalyst. The Franco-Prussian War of 1871 proved inspirational for publisher Edward Stanford, who produced this cartographic take on the situation in France at the time of the Prussian invasion, using art as propoganda. A creature, part-human, part-monster, extends westwards from Germany, its fangs coming to rest above Tours. The map's user is being induced to adopt an anti-Prussian stance. Prussia's aggression is to be abhorred, yet no wrongdoing is suggested on the part of France. The question to ask is whether this genre of mapping is designed to educate or to entertain, or both?

Edward Stanford Ltd, *Prussia Pausing*, 1871.
C21 (110)

14th Feb 1871

PRUSSIA PAUSING.

OR The ACCURATE ARMISTICE DEMARCATION LINE.

The Outlines of the Animal denote with ABSOLUTE ACCURACY, the lines held by the Belligerents during the Armistice as specified in the full text of the Convention, and the Area of French Territory occupied by the Germans at the date of the FALL OF PARIS.

Attention is drawn to the extraordinary coincidence of the Armistice boundaries representing the outlines of a carnivorous animal typical of the relentless voracity of Prussia, as evinced by the rapacity of her present demands.

PUBLISHED BY EDWARD STANFORD, 6, CHARING CROSS, LONDON. 14TH FEB. 1871.

Ent: Sta: Hall.

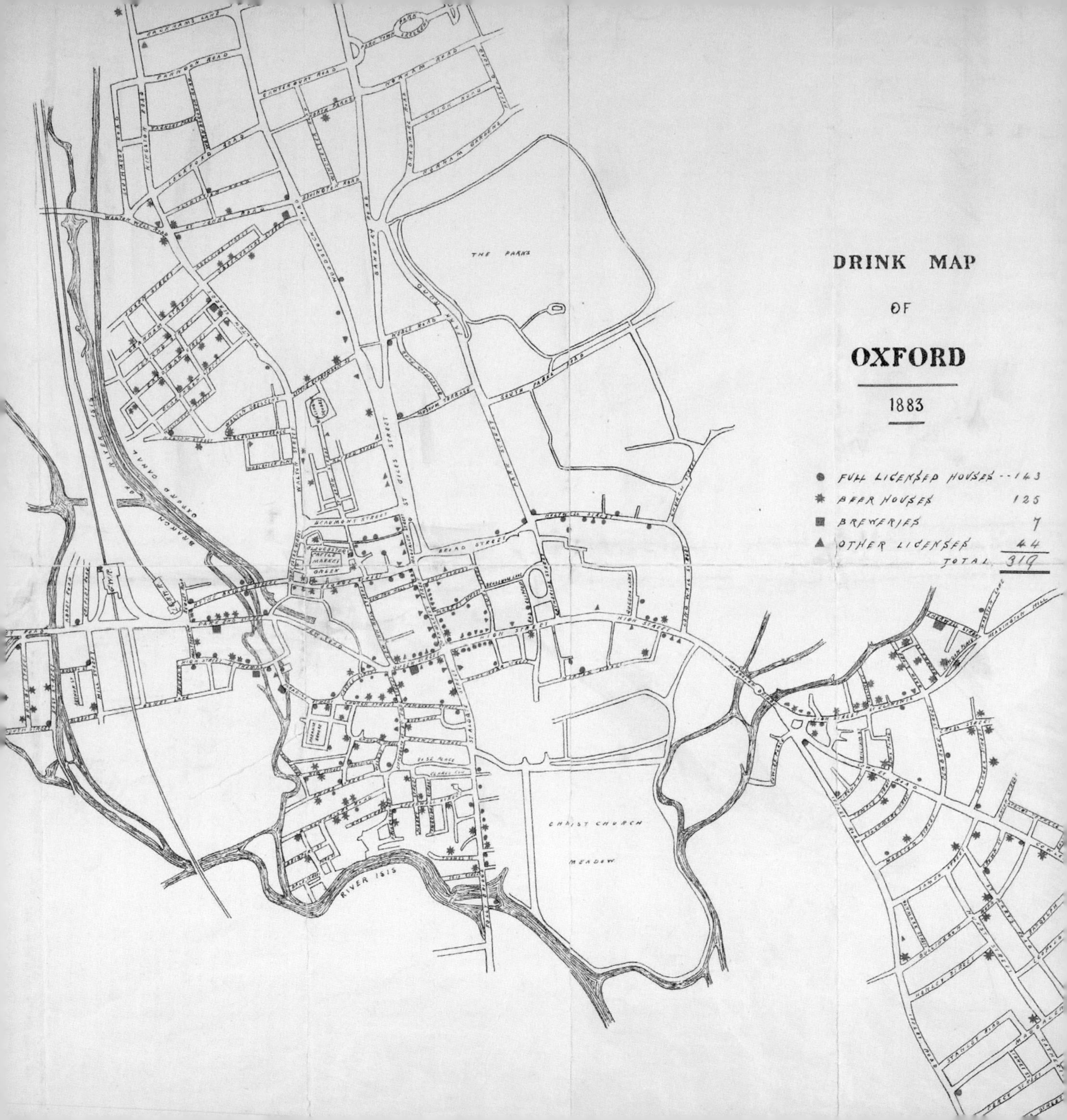
DRINK MAP
OF
OXFORD
1883
FULL LICENSED HOUSES --143
BEER HOUSES 125
BREWERIES 7
OTHER LICENSES 44
TOTAL 319
THE PARKS
CHRIST CHURCH
MEADOW
RIVER ISIS
BROAD STREET
HIGH STREET
BEAUMONT STREET
ST GILES STREET
PARK STREET
SOUTH PARK ROAD
BANBURY ROAD
WOODSTOCK ROAD
NORHAM ROAD
NORHAM GARDENS
CANTERBURY ROAD
KEBLE ROAD
WALTON STREET
OXFORD CANAL

30

WHERE TO BUY A DRINK IN OXFORD

The Temperance Union of Oxfordshire's 'Drink Map' is heavy on message and effective in design, its striking use of red symbols bringing the map to life. Each represents one of the 319 drinking establishments in Oxford, divided into four categories. The map's reverse states that every twenty-second house is a 'drink shop', and this figure is 50 per cent higher than the national average. There are red clusters in Jericho (northwest of the centre), St Ebbe's (southwest) and St Clement's (east), all areas noted as housing those with lowest incomes. Predominantly wealthy North Oxford is a relative desert, as it remains in the twenty-first century.

Drink Map of Oxford, 1883. C17:70 Oxford (7)

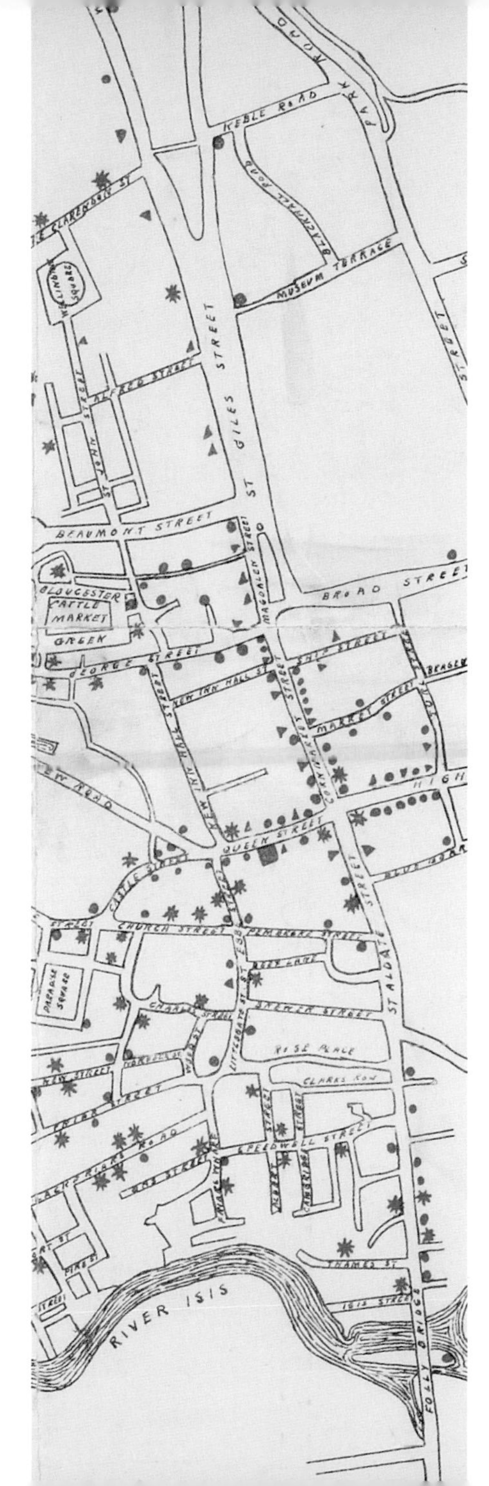

31

TREASURE ISLAND

Probably the most iconic of all fictional maps is Robert Louis Stevenson's 'Treasure Island'. It was drawn to accompany his book of the same name, published in 1883: a tale of piracy and lost riches, the escapades of Jim Hawkins and Long John Silver, and at its heart Billy Bones's treasure map. Stevenson drew the map in an idle moment in 1881, realising it 'was the chief part of my plot'. The story quickly followed. He claims the original map was subsequently lost, but insisted on its enduring importance to his story: 'the tale has a root there; it grows in that soil'.

Robert Louis Stevenson, 'Treasure Island' map, 1883. Arch. AA e.149

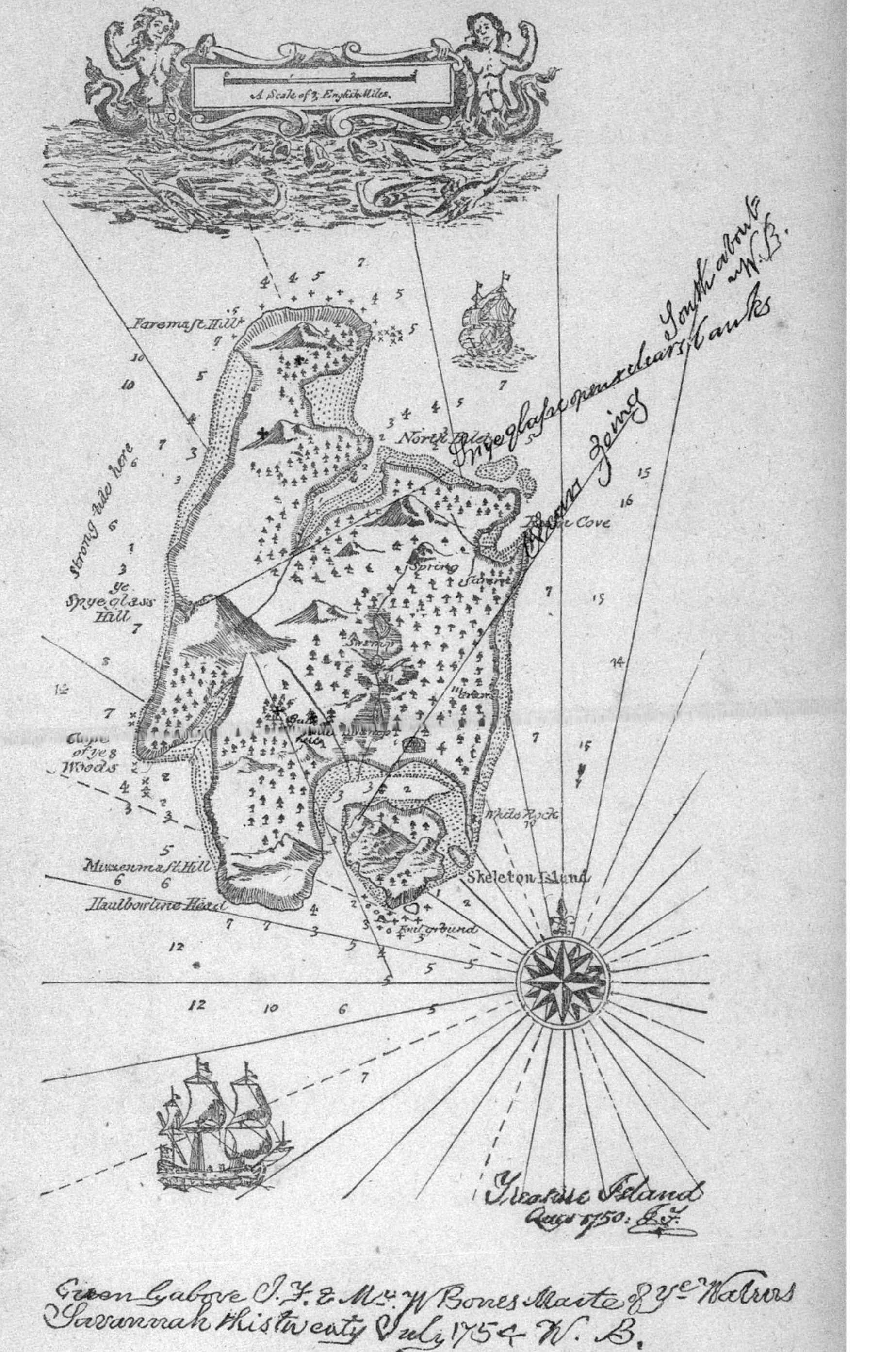

Facsimile of Chart; latitude and longitude struck out by J. Hawkins

32

A PACIFIC STICK CHART

Sailors from the Marshall Islands in the Pacific Ocean developed a unique mapping method to enable them to sail from one island to another. So-called 'stick charts' were made from wood, cane and coconut fronds tied together, with shells or knots to designate islands. Each stick represents the 'swells', or surface gravity waves, unique to the Marshall Islands, which were used to navigate between sea and land. Marshallese pilots – called *ri-metos*, or 'wave pilots' – would memorize such stick charts before sailing in their canoes. European colonists began collecting such charts in the nineteenth century, although they are probably much older.

Marshall Islands stick chart, 1896. Pitt Rivers Museum 1897.1.2

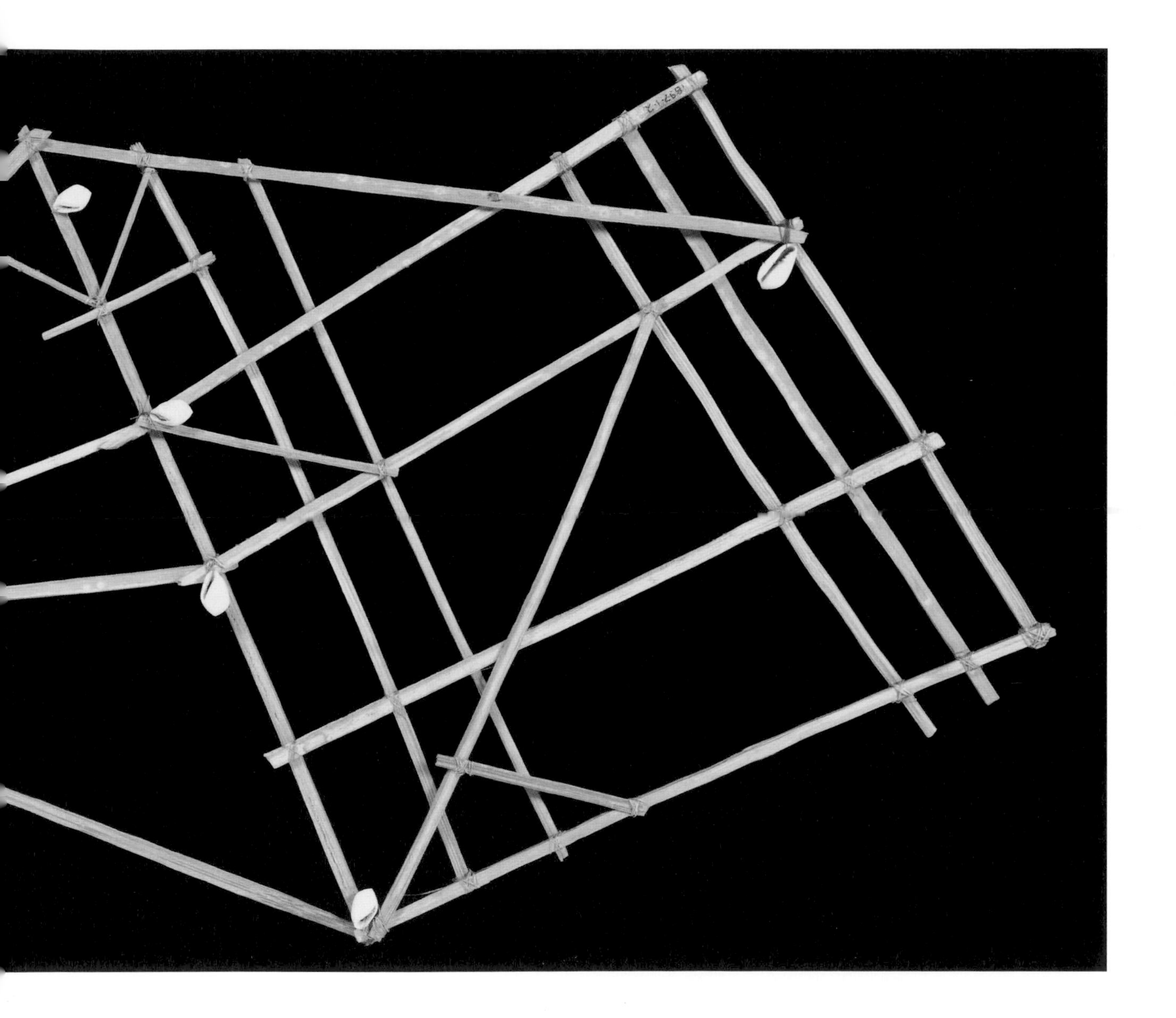
1897.1.2

33

THE TIBETAN WHEEL OF LIFE

The complex cosmology of Tibetan Buddhism and its central belief in *saṃsāra* – the cycle of transmigration and rebirth – in the soul's quest for release (*nirvana*) gave rise to extraordinarily elaborate maps such as this *Bhavacakra*, or 'Wheel of Life/Becoming' (*srid pa'i 'khor lo* in its Tibetan transliteration), most like this one painted on silk or cotton hangings called *thangkas*. Here the realm of desire (*Kāmadhātu*) is embraced by the wrathful deity Shinjé. The faithful can reflect on the soul's transmigration through the infernal realms, driven by the defining emotions of desire, anger and ignorance personified by the cock, snake and pig.

Tibetan *thangka* of the *Bhavacakra*, late nineteenth century, Ashmolean Museum, EA 1956.185

34

A MAP OF THE BRITISH EMPIRE

The height of the British Empire at the beginning of the twentieth century saw an outpouring of maps and atlases for use in government, education and trade. Philip's was one of London's oldest publishing houses, famed for its atlases and educational maps, such as this one, which were widely distributed at home and abroad. Drawn on the Mercator projection, this one shows British imperial possessions coloured in shades of red, as well as dozens of maritime routes running east to west that defined the commercial extent of the empire, each carefully measured in miles. Trade is seen here as the driver of empire.

Philip's Mercantile Chart of the Commercial Routes to the East, 1906. B1 (244)

MERCANTILE CHART OF THE COMMERCIAL ROUTES TO THE EAST

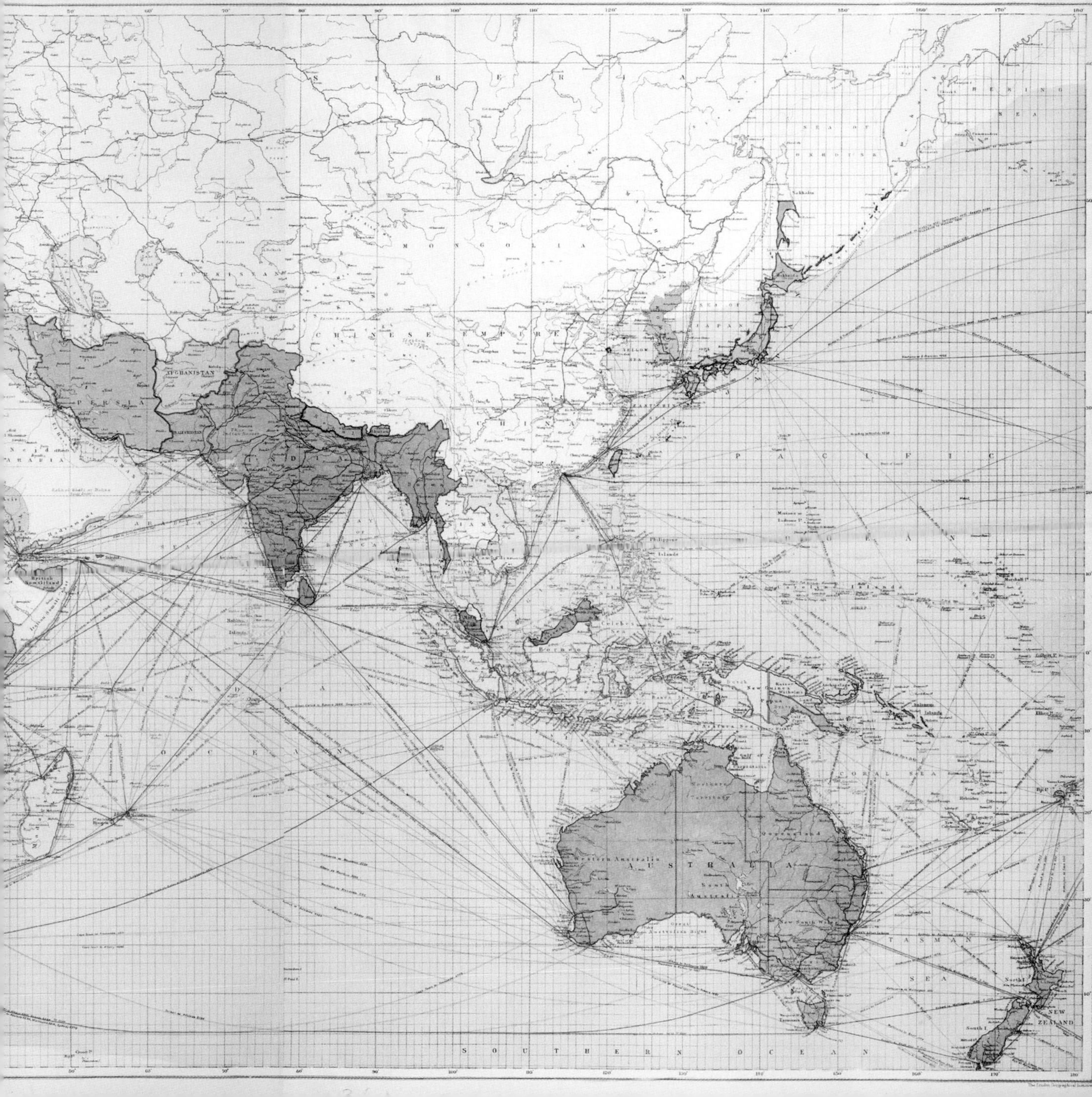

S I B E R I A

MONGOLIA

TURKISTAN

CHINESE EMPIRE

TIBET

CHINA

YELLOW SEA

SEA OF JAPAN

JAPAN

SEA OF OKHOTSK

BERING SEA

Sakhalin

AFGHANISTAN

PERSIA

BALUCHISTAN

ARABIA

INDIA

SIAM

Philippine Islands

British Somaliland

Italian Somaliland

P A C I F I C

O C E A N

Marshall Is.

Carolines Islands

Borneo

Celebes

New Guinea

Bismarck Archipelago

Solomon Islands

Gilbert Is.

Ellice Is.

I N D I A N

O C E A N

CORAL SEA

New Hebrides

Fiji Is.

Northern Territory

Queensland

Western Australia

AUSTRALIA

South Australia

New South Wales

Great Australian Bight

TASMAN SEA

North I.

South I.

NEW ZEALAND

S O U T H E R N   O C E A N

50° 60° 70° 80° 90° 100° 110° 120° 130° 140° 150° 160° 170° 180°

35

HMS *CHALLENGER*'S MAP OF THE PACIFIC OCEAN

By the nineteenth century steam-driven maritime exploration was directed by scientific interest in what lay beneath the oceans. Bathymetric maps of the ocean floor like this one of the Pacific were made during the HMS *Challenger* expedition (1862–6). It covered 110,867km, stopping at 362 stations and taking more than 100 bottom dredges of the ocean floor. In March 1875 the crew recorded a sounding of 4,475 fathoms (8,184m) in the southwest Pacific. Subsequently recorded as Challenger Deep, it remains the deepest known point on the Earth's seabed. The *Challenger*'s methods for the collection and collation of bathymetric data laid the foundation for modern oceanography.

Bathymetric map of the Pacific Ocean showing the route of HMS *Challenger*, [1914], J1 (181)

Voyage of H.M.S. "Challenger"
Summary of Results Chart I B
BATHYMETRICAL CHART
OF THE
PACIFIC OCEAN
North Pole
ALEUTIAN BASIN
OKHOTSK BASIN
MAURY DEEP
JAPAN BASIN
TUSCARORA DEEP
MURRAY DEEP
CLOVER DEEP
TANNER DEEP
HAWAIIAN PLATEAU
Tropic of Cancer
WYMAN DEEP
AMMEN DEEP
RENARD DEEP
CALIFORNIAN PLATEAU
MEXICO BASIN
CARIBBEAN BASIN
CHINA BASIN
SWIRE DEEP
PHILLPPINE BASIN
LADRONE PLATEAU
BROOKE DEEP
BELKNAP DEEP
CHALLENGER DEEP
SULU BASIN
CAROLINE PLATEAU
MARSHALL PLATEAU
GREY DEEP
CELEBES BASIN
CAMPBELL DEEP
GALAPAGOS PLATEAU
MILLER DEEP
Equator
BANDA BASIN
SOLOMON ISLANDS PLATEAU
HILGARD DEEP
MARQUESAS PLATEAU
CARPENTER BASIN
AGASSIZ BASIN
FIJI PLATEAU
OLDHAM DEEP
PAUMOTU PLATEAU
MILNE-EDWARDS DEEP
WHARTON BASIN
ALBATROSS PLATEAU
Tropic of Capricorn
GAZELLE BASIN
BUCHAN BASIN
NEW ZEALAND PLATEAU
ALDRICH DEEP
THOMSON BASIN
HAECKEL DEEP
JEFFREY DEEP
JUAN FERNANDEZ PLATEAU
TASMANIAN PLATEAU
BAKER BASIN
Antarctic Circle
South Pole
NOTE
Extreme Limit of Drift Ice
Average Limit of Drift Ice
Limits of Pack Ice
BASIN is applied to areas encircled by low submarine ridges
DEEP with over 3000 fathoms
All Ocean Soundings on the chart over 1000 fathoms are shown by the first two figures,
the last two figures being always omitted, thus 4475 fathoms is represented by 44
For explanation of abbreviations see Appendix I.
REFERENCE TO COLOURING OF CHART *
CONTOURS SHOWING HEIGHT OF LAND
CONTOURS SHOWING DEPTH OF SEA
SEA LEVEL
Feet
Fathoms
* The Unexplored Region is left White
Andrew J. Herbertson.
J. G. Bartholomew.

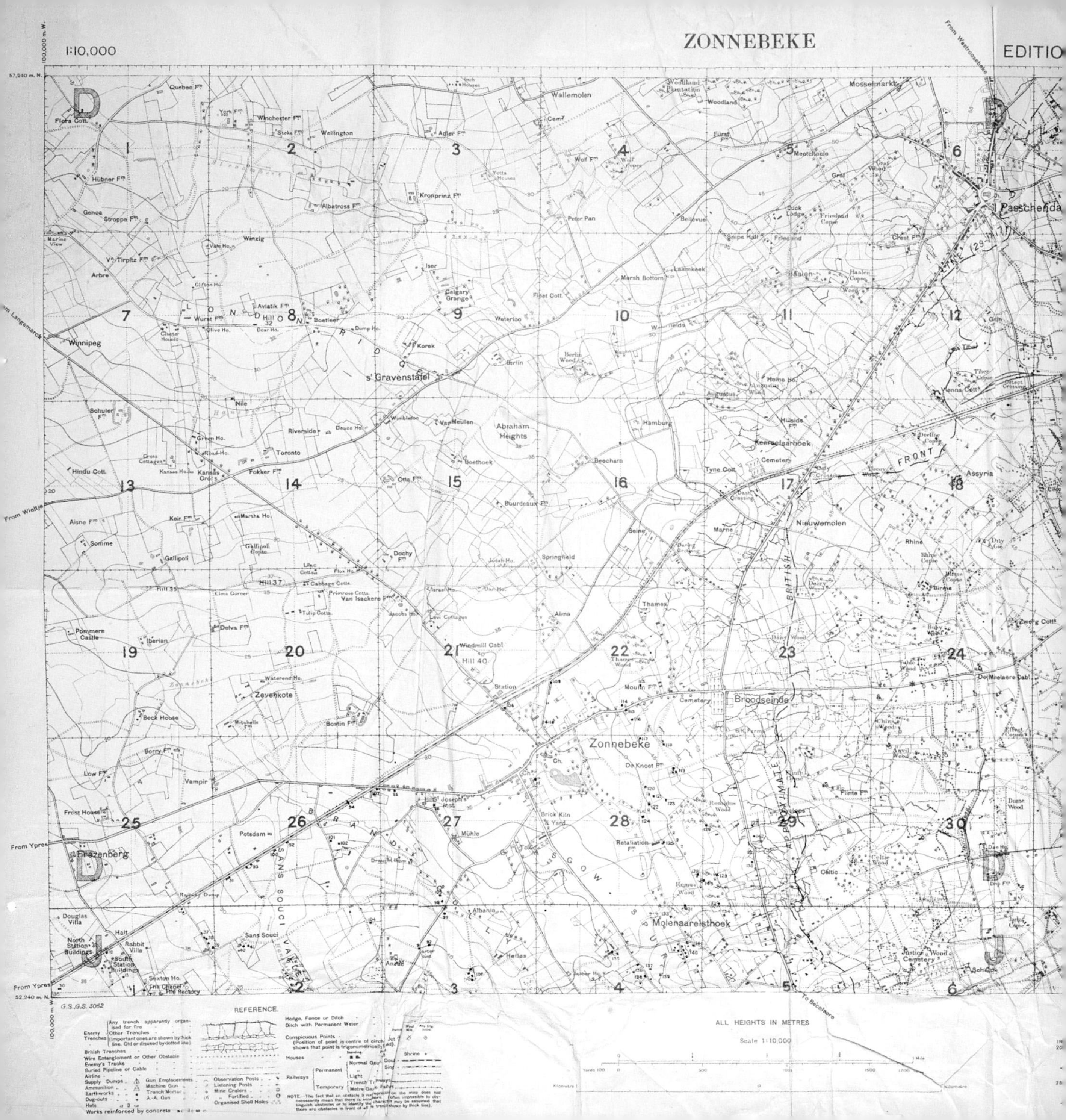
1:10,000
ZONNEBEKE
EDITIO
From Westroosebeke
Quebec Fm
Flora Cott.
York Fm
Winchester Fm
Wellington
Adler Fm
Wallemolen
Woodland Plantation
Woodland
Mosselmarkt
Hübner Fm
Kronprinz Fm
Albatross Fm
Genoa
Stroppe Fm
Yetta Houses
Peter Pan
Wolf Fm
Wolf Copse
Meetcheele
Bellevue
Duck Lodge
Friesland Copse
Passchenda
Marine View
Vn Tirpitz Fm
Winzig
Vale Ho.
Iser
Snipe Hall
Friesland
Crest Fm
Arbre
Marsh Bottom
Laamkeek
Haalen
Haalen Copse
Calgary Grange
Aviatik Fm
Wurst Fm
Hill 32
Boetleer
Clifton Ho.
Olive Ho.
Dear Ho.
Dump Ho.
Chalet Houses
Waterloo
Berlin Wood
Berlin
Korek
s' Gravenstafel
Winnipeg
From Langemarck
Heine Ho.
Augustus
Augustus Wood
Tiber
Tiber Copse
Vienna Cott.
Nile
Schuler Fm
Wimbledon
Van Meulen
Abraham Heights
Hamburg
Hillside Fm
Keerselaarhoek
Riverside
Deuce Ho.
Green Ho.
Road Ho.
Toronto
Cross Cottages
Kansas Ho.
Kansas Cross
Hindu Cott.
Fokker Fm
Boethoek
Beecham
Tyne Cott
Cemetery
Otto Fm
Assyria
Bourdeaux Fm
Nieuwemolen
Marne
From Wieltje
Aisne Fm
Somme
Keir Fm
Martha Ho.
Gallipoli
Gallipoli Copse
Dochy Fm
Seine
Springfield
Rhine
Lilac Cotts.
Flox Ho.
Hill 35
Hill 37
Cabbage Cotts.
Elms Corner
Primrose Cotts.
Van Isackere Fm
Thames
Tulip Cotts.
Jacobs Ho.
Levi Cottages
Alma
Pommern Castle
Iberian
Delva Fm
Windmill Cabt
Hill 40
Dairy Wood
Zwarg Cott.
Zonnebeke
Waterend Ho.
Zevenkote
Station
Moulin Fm
Broodseinde
Cemetery
De Mnelaere Cabt
Beck House
Mitchells Fm
Bostin Fm
Borry Fm
De Knoet Fm
Anvil Wood
Low Fm
Vampir
Flinte Fm
Frost House
St Joseph's Inst.
Brick Kiln Yard
Potsdam
Mühle
Retaliation Fm
Romulus Wood
Dame Wood
From Ypres
Frezenberg
Railway Dump
Remus Wood
Celtic Wood
Celtic
Douglas Villa
Halt
Albania
Molenaarelsthoek
North Station Buildings
Rabbit Villa
Sans Souci
South Station Buildings
Helles
Anzac
Justice Wood
Cemetery
Sexton Ho.
The Chapel
The Rectory
To Becelaere
SANS SOUCI VALLEE
LONDON RIDGE
BRITISH FRONT LINE (29-11-17)
APPROXIMATE BRITISH FRONT
BRANDY GULLY
GLASGOW SPUR
1 2 3 4 5 6 7 8 9 10 11 12 13 14 15 16 17 18 19 20 21 22 23 24 25 26 27 28 29 30
57,240 m. N.
52,240 m. N.
100,000 m. W.
G.S.G.S. 3062
REFERENCE.
Enemy Trenches
Any trench apparently organised for fire
Other Trenches
(Important ones are shown by thick line. Old or disused by dotted line)
British Trenches
Wire Entanglement or Other Obstacle
Enemy's Tracks
Buried Pipeline or Cable
Airline
Supply Dumps
Ammunition
Earthworks
Dug-outs
Huts
Works reinforced by concrete
Gun Emplacements
Machine Gun
Trench Mortar
A.-A. Gun
Observation Posts
Listening Posts
Mine Craters
Fortified
Organised Shell Holes
Hedge, Fence or Ditch
Ditch with Permanent Water
Conspicuous Points
Houses
Shrine
Railways
Permanent
Normal Gauge
Light
Temporary
Trench Tramways
Metre Gauge
ALL HEIGHTS IN METRES
Scale 1:10,000
Yards
Kilometre
Mile

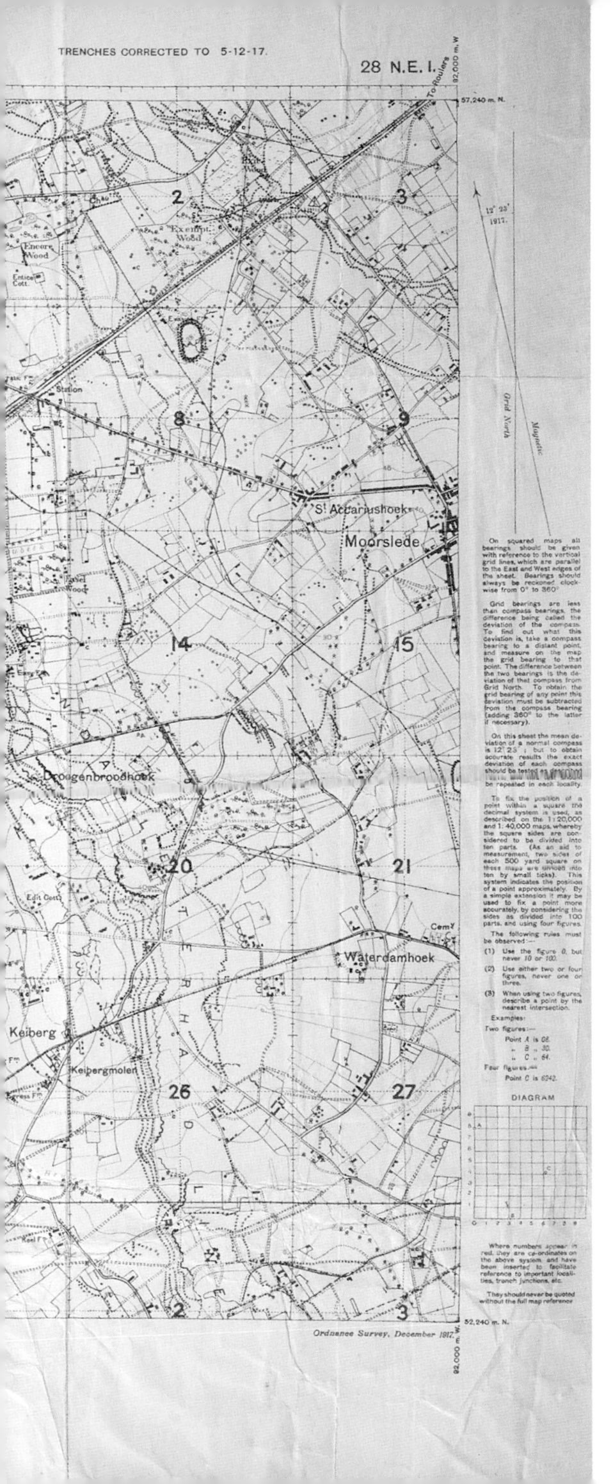

36

BATTLE OF YPRES TRENCH MAP

This trench map, issued to soldiers on the front line, includes Passchendaele in northwestern Belgium, scene of the Third Battle of Ypres. The area saw heavy fighting from July to November 1917 as the Allies sought to gain control of the ridges east of Ypres. Derived from existing Belgian mapping, this sheet would have been cartographically enhanced thanks to aerial photography. German trenches are indicated in red in considerable detail. Less minutely represented are those of the British, marked in blue behind a heavier pecked blue line showing 'Approximate British front line 29-11-17'. This is edition 9A, numbering which shows how frequently trench maps were updated.

Ordnance Survey/Geographical Section General Staff map of Zonnebeke, 1:10,000, sheet 28 N.E. I, 1917. C1 (3) 720

37

SUBURBAN APARTHEID IN OXFORD

Cutteslowe is typical of inter-war suburbia, but with a dark past which delivered small-scale economic apartheid to Oxford. Walls were built across two streets to prevent tenants from council housing constructed to the east from passing by the neighbouring private development of more expensive houses built a few years earlier. Those walls remained for twenty-five years until their demolition in 1959. This map shows both estates in pink and yellow, the north and south walls in red. Walking distances from either side of the wall to bus stops and schools are highlighted. One bus stop was a 385-yard walk from the west side of the South Wall, but 1,075 yards from the east.

Oxford City Council plan of Cutteslowe, showing the North Wall and the South Wall. J.F. Richardson, 1934

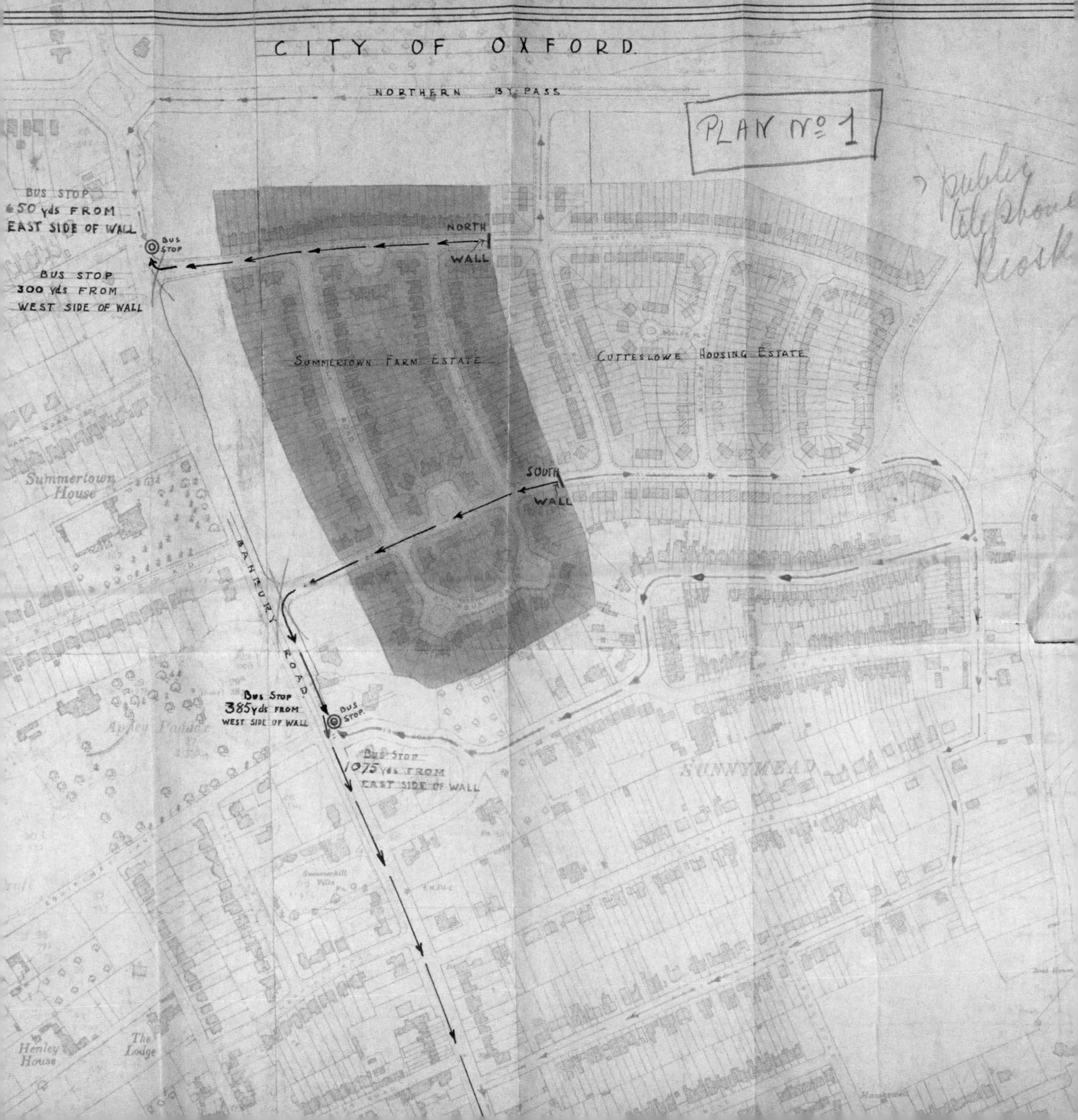

CITY OF OXFORD.

NORTHERN BY-PASS

PLAN No 1

public telephone kiosk

BUS STOP
650 yds FROM
EAST SIDE OF WALL

BUS STOP

BUS STOP
300 yds FROM
WEST SIDE OF WALL

NORTH
WALL

SUMMERTOWN FARM ESTATE

CUTTESLOWE HOUSING ESTATE

SOUTH
WALL

Summertown
House

BANBURY ROAD

Bus Stop
385yds FROM
WEST SIDE OF WALL

BUS STOP.

Bus Stop
1075yds FROM
EAST SIDE OF WALL

SUNNYMEAD

Summerhill
Villa

Henley
House

The
Lodge

38

MAPPING TOLKIEN'S MIDDLE-EARTH

When writing *The Hobbit* (1937) and *The Lord of the Rings* (1954–5), J.R.R. Tolkien insisted on the need to 'first make a map and make the narrative agree'. In the 1930s he drew maps describing his 'legendarium' (fictional mythology). In this small-scale world map Tolkien depicts the moment when Melkor (subsequently Morgoth, the 'Dark Enemy'), the most powerful Valar, destroys the world's vast icy lamps, plunging it into darkness, and creating the seas of Helkar and Ringil. Melkor built the Northern Towers, or Iron Mountains, and the fortress of Utumna while the Valar retreated westwards to Valinor. Middle earth (in capital letters and not yet hyphenated) is shown for the very first time.

J.R.R. Tolkien, 'The world about V.Y. 500 after the fall of the lamps', 1930s. MS. Tolkien S 2/III, fol. 7r

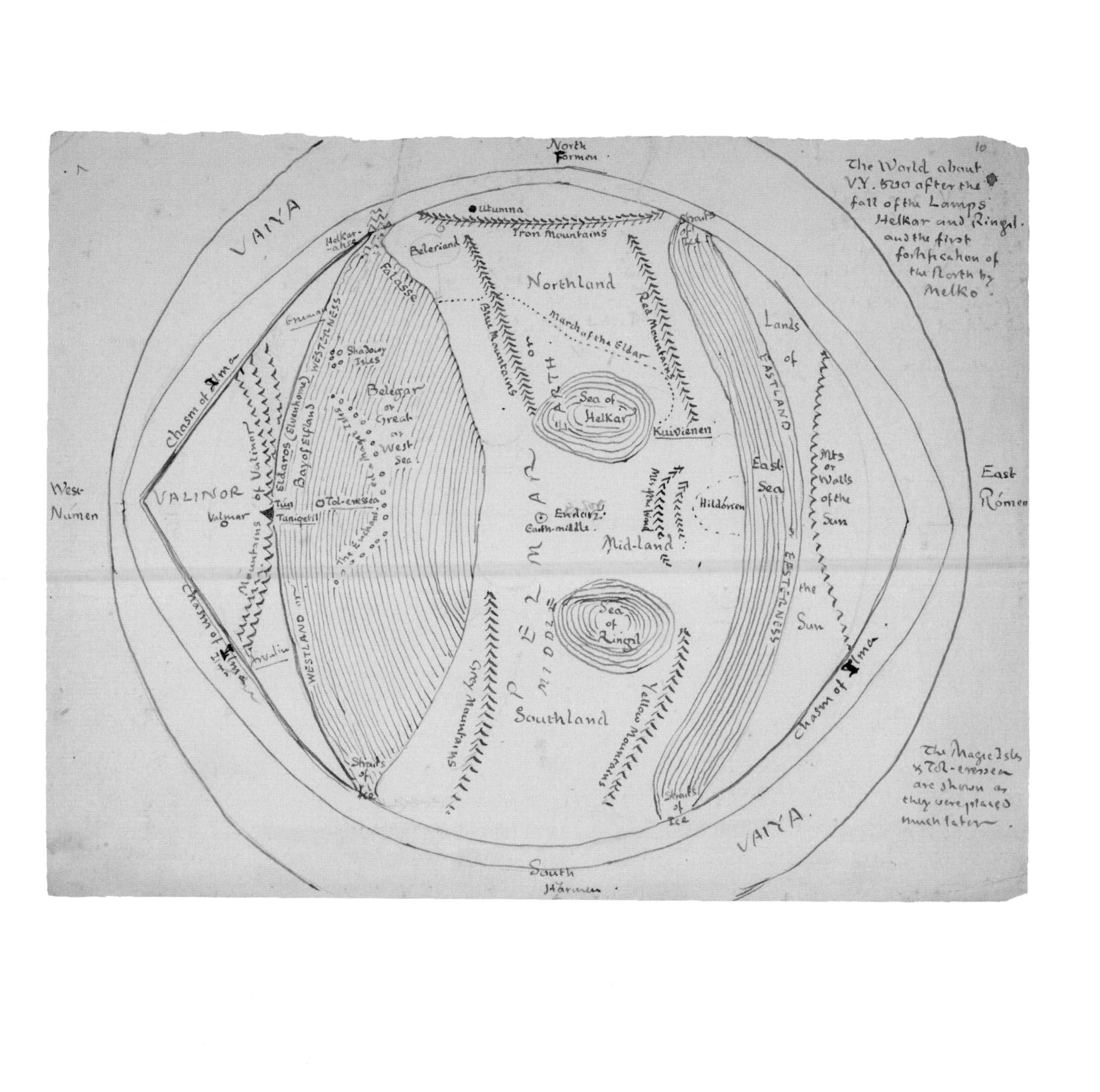
10
The World about V.Y. 500 after the fall of the Lamps Helkar and Ringil. and the first fortification of the North by Melko.
North
Formen
VAIYA
Utumna
Iron Mountains
Beleriand
Northland
Falasse
Blue Mountains
March of the Eldar
Red Mountains
Lands of the Sun
EASTLAND
Shadowy Isles
Belegar or Great or West Sea
Sea of Helkar
Kuivienen
Chasm of Ilma
Mountains of Valinor
Eldaros (Elvenhome)
Bay of Elfland
WESTERNESS
West
Númen
VALINOR
Valmar
Tún
Taniqetil
Tol-eressea
Mts or Walls of the Sun
East Sea
East
Rómen
Endor
Earth-middle
Hildórien
Mid-land
EASTERNESS
the Sun
Sea of Ringil
MIDDLE EARTH
WESTLAND
Grey Mountains
Yellow Mountains
Southland
Straits of Ice
The Magic Isles & Tol-eressea are shown as they were placed much later.
South
Harmen

HELM'S DEEP
& the HORNBURG

nded his letters for publication and condemned
affectation & self-consciousness of Pope's letters.
Cowper describes the arrival of a sundial, the
rgement of the greenhouse, where he often worked in
mmer, – the history of the card table, the character of
ome hare. All these things come alive & interest
ance of Cowper's charm & whimsical sensitive
ke. It is his personality that gives charm both to his
a well as his letters. He is not great or sublime
will always [illegible] because of his gentle & gay
[illegible]

TOLKIEN'S MAP OF HELM'S DEEP AND THE HORNBURG

Tolkien turned to larger-scale maps to envisage key moments in his books. On an Oxford examination paper he was marking in 1942, he drew maps of Helm's Deep valley and the Hornburg fortress, the site of the climactic battle between the armies of Saruman and the Rohan under King Théoden in *The Two Towers*, the second volume of *The Lord of the Rings*. Drawn from an aerial perspective, with a compass rose, scale bar and stippled routes showing the movement of the armies, it also shows Tolkien drafting his narrative with a written exchange between the fortress's defenders at the top.

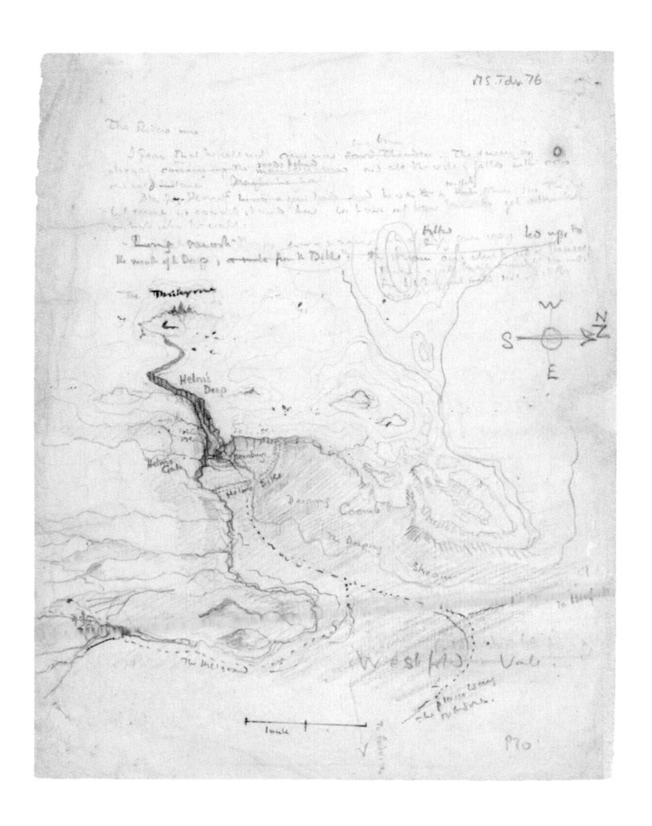

J.R.R. Tolkien, 'Helm's Deep & the Hornburg', *c.*1942. MS. Tolkien Drawings 76r

England 1:50.000

Achtung!
Mit deutschem Gauß-Krüger-Gitternetz im 0. Streifen
Es dürfen zur Berechnung nur Werte ein und desselben Streifens verwendet werden!

Das deutsche Gauß-Krüger-Gitternetz ist in Abständen von 2 km in Rot durchgezogen.

CLACTON ON SEA & HARWICH

Alle Höhenangaben au…
angegeben. Zur groben…
Alle Höhenangaben de…
Meter-Angaben zu er…

WEST SUFFOLK
EAST SUFFOLK
MANNINGTREE
COLCHESTER
Great Bromley
Little Bentley
BRIGHTLINGSEA
Little Clacton
CLACTON ON SEA
Mersea Flats
RIVER BLACKWATER
Wallet

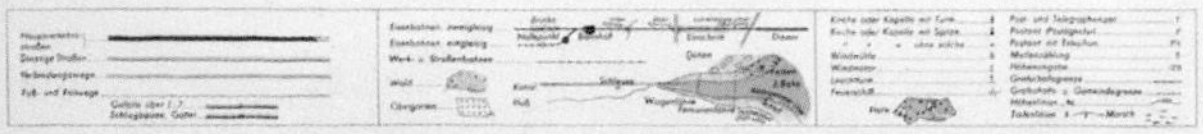

Maßstab 1:50.000

Höhenangaben in engl. Fuß über mittl. Meeresspiegel. Abstand der Höhenlinien 50 engl. Fuß (= ca 15 m)
Tiefenangaben in englischen Faden bezogen auf mittleres Springniedrigwasser
Abstand der Tiefenlinien 5 engl. Faden (= ca 9 m)

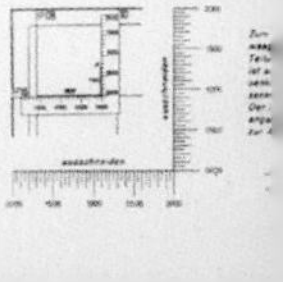

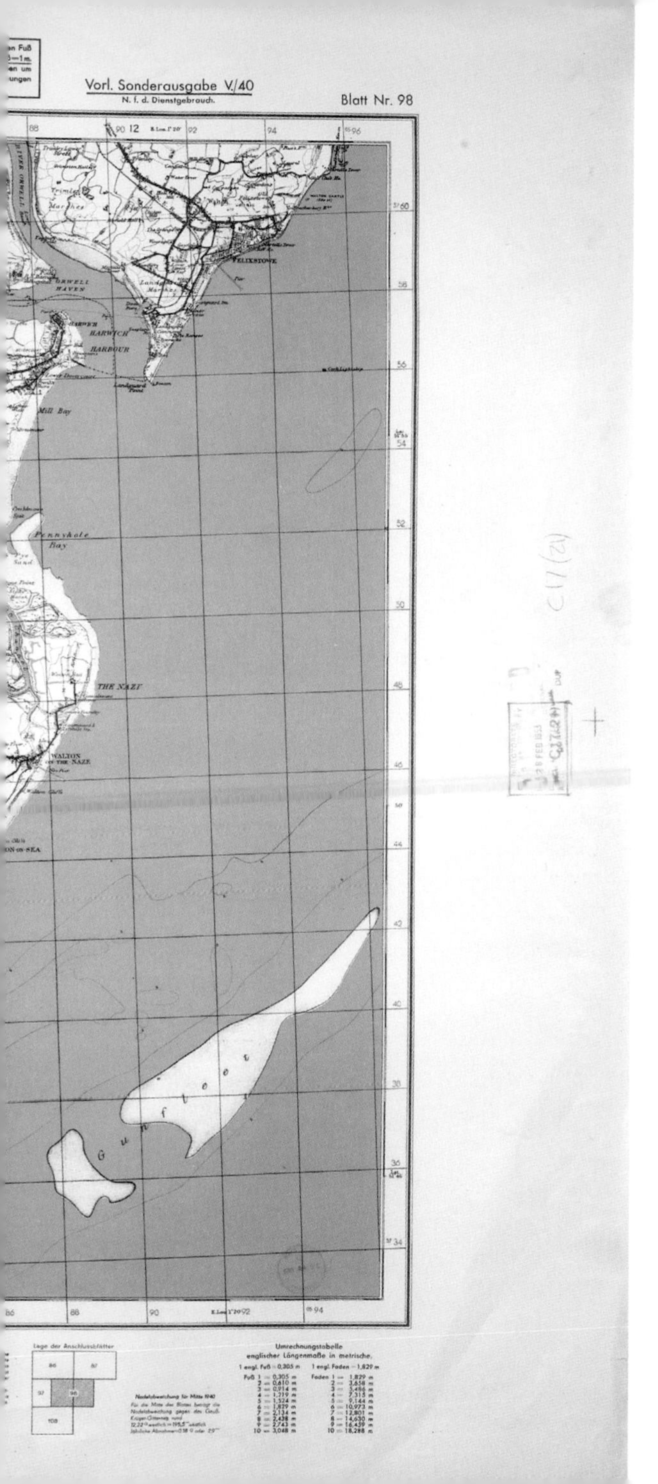

39

SECOND WORLD WAR RECYCLED MAP

This map was initially published by Ordnance Survey in 1921 as sheet 98 of the Popular Edition One-Inch Map series, 'Clacton on Sea & Harwich'. With the onset of war imminent, copies of the map were acquired by German agents to prepare for an invasion. The German military map-makers enlarged the Popular Edition to a more familiar 1:50,000 scale and overlaid a new grid, explaining its offset nature. A new German series 'England 1:50 000' was created, and the maps dated 1940. The surrounding marginalia, with explanatory German text, details how to interpret the 'English foot', aiming to bring sense to this confusing unit of measurement.

Generalstab des Heeres, England 1:50 000, sheet 98, Clacton on Sea & Harwich, 1940. C17 (21)

40

D-DAY LANDING MAP

'Sword Area' shows preparations for the D-Day landings in June 1944, covering an area of 7 x 7 km. It is one of a set of five sheets. Sword Beach was 8 km wide, and this sheet shows its eastern flank. Peter, Queen and Roger (seen here in bold) were codenames identifying sectors of beach, named alphabetically from west to east, themselves subdivided into Green, White and Red. The eventual landing focused on the 3 km-wide strip designated White and Red in Queen sector. Purple text shows that the map's information was current on 6 April 1944, two months ahead of the invasion; it warns that underwater obstacles were being added – a worrying disclaimer of reliability.

'Operation Neptune 1', 1:12,500, 'Sword Area', 1944.
C21:37 (28)

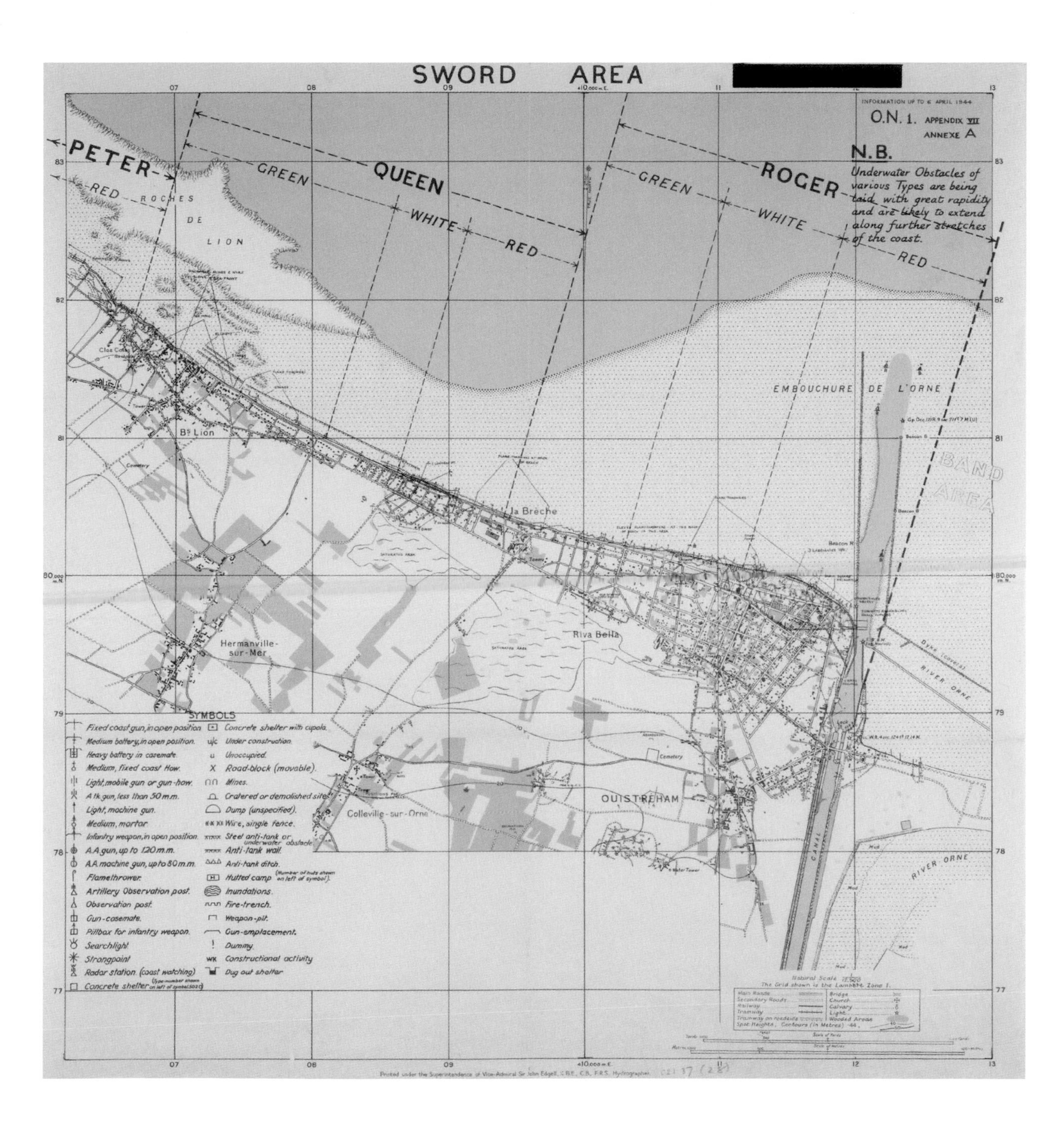
SWORD AREA
INFORMATION UP TO 6 APRIL 1944
O.N. 1. APPENDIX VII
ANNEXE A
N.B.
Underwater Obstacles of various Types are being laid with great rapidity and are likely to extend along further stretches of the coast.
PETER
QUEEN
ROGER
GREEN
WHITE
RED
ROCHES DE LION
EMBOUCHURE DE L'ORNE
BAND AREA
Bg Lion
La Brèche
Riva Bella
Hermanville-sur-Mer
Colleville-sur-Orne
OUISTREHAM
Cemetery
CANAL
RIVER ORNE
Dyke (covers)
SYMBOLS
Fixed coast gun, in open position
Medium battery, in open position.
Heavy battery in casemate.
Medium, fixed coast How.
Light, mobile gun or gun-how.
A tk.gun, less than 50 m.m.
Light, machine gun.
Medium, mortar.
Infantry weapon, in open position.
A.A.gun, up to 120 m.m.
A.A.machine gun, up to 50 m.m.
Flamethrower.
Artillery Observation post.
Observation post.
Gun-casemate.
Pillbox for infantry weapon.
Searchlight.
Strongpoint.
Radar station (coast watching)
Concrete shelter
Concrete shelter with cupola.
Under construction.
Unoccupied.
Road-block (movable).
Mines.
Cratered or demolished site.
Dump (unspecified).
Wire, single fence.
Steel anti-tank or underwater obstacle.
Anti-tank wall.
Anti-tank ditch.
Hutted camp
Inundations.
Fire-trench.
Weapon-pit.
Gun-emplacement.
Dummy.
Constructional activity
Dug out shelter
Printed under the Superintendence of Vice-Admiral Sir John Edgell, C.B.E., C.B., F.R.S., Hydrographer.

41

NOW YOU SEE IT, NOW YOU DON'T

In the late 1940s the Royal Air Force was commissioned to photograph large swathes of Britain, resulting in a series of black-and-white 'photomosaics' at a scale of 1:10,560. The photographs were enhanced with overprinted place names. The objective was to aid post-war reconstruction, though they were also made available for general sale. This image is centred on Colnbrook, just west of Heathrow, with the geographical extent of the image measuring 5 x 5 km. This area is now much changed, with the M4 running east–west across the centre, the M25 north–south along its eastern edge, and the Queen Mother Reservoir located southwest of Colnbrook.

Ordnance Survey, 'Air Photo Mosaic' centred on Colnbrook, sheet 51/07 N.W., *A* edition, 1948

ORDNANCE SURVEY
UCKINGHAMSHIRE
51/07 N.W.
RICHINGS PARK
COLNBROOK
HORTON

The authorities became aware that photomosaics were being purchased by foreign individuals and grew concerned about the security implications, as the maps showed everything visible on the ground. Editing resulted in the publication of *B* editions, immediately indicating that something(s) on the initial edition should not have been publicised. There is a subtle alteration towards the top centre of this revised map, where open land with industrial buildings just south of the railway, and a larger complex of buildings further south, could be seen on the first edition. This has disappeared under a pattern of fields, with new roads inserted to conceal the Hawker Aircraft factory, airfield and parked aeroplanes.

Ordnance Survey, 'Air Photo Mosaic' centred on Colnbrook, sheet 51/07 N.W., *B* edition, 1948

ORDNANCE SURVEY
UCKINGHAMSHIRE
51/07 N.W.
RICHINGS PARK
HORSEMOOR GREEN
MIDDLESEX
COLNBROOK
POYLE
HORTON

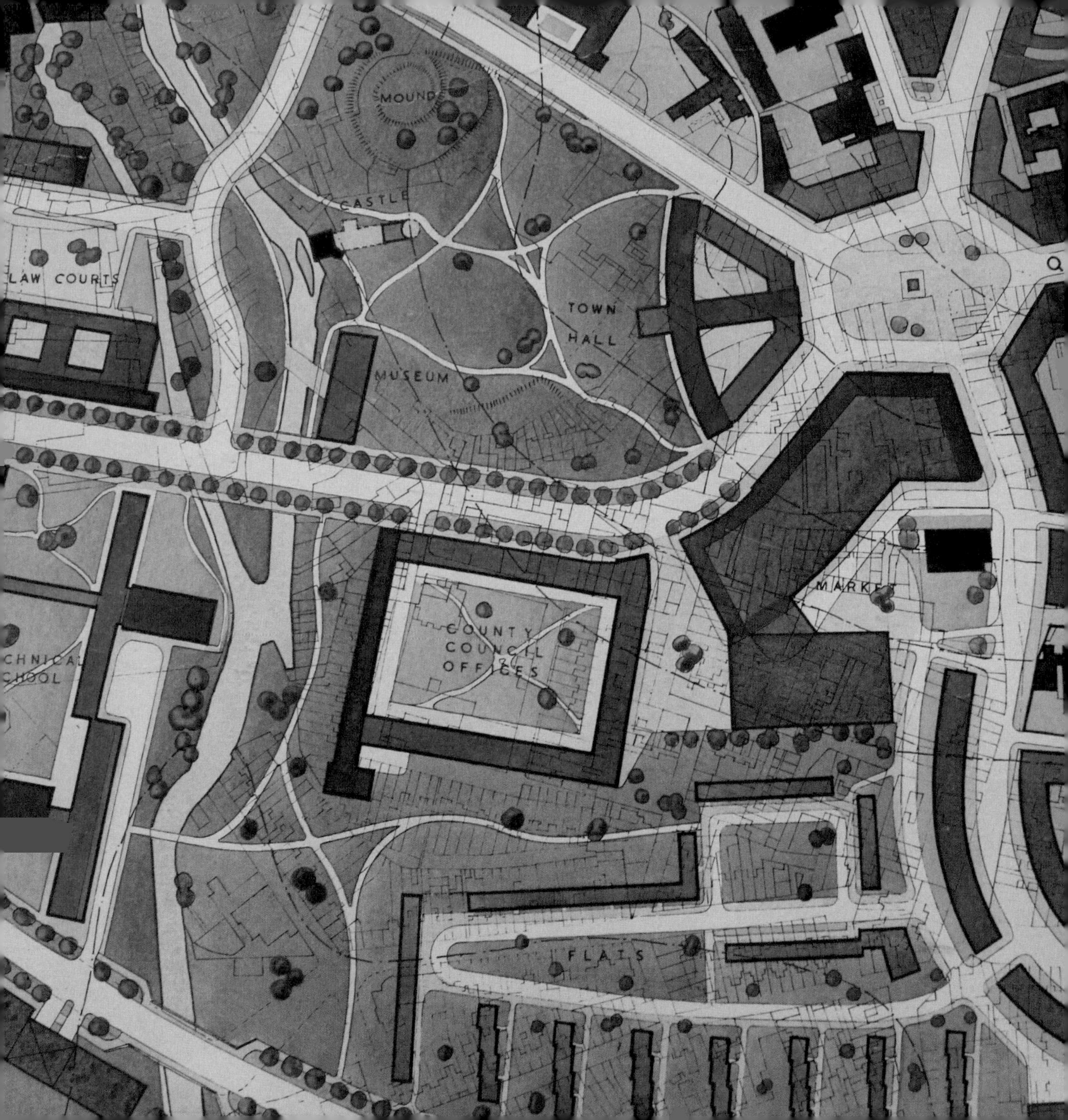
MOUND
CASTLE
LAW COURTS
TOWN HALL
MUSEUM
MARKET
COUNTY COUNCIL OFFICES
FLATS

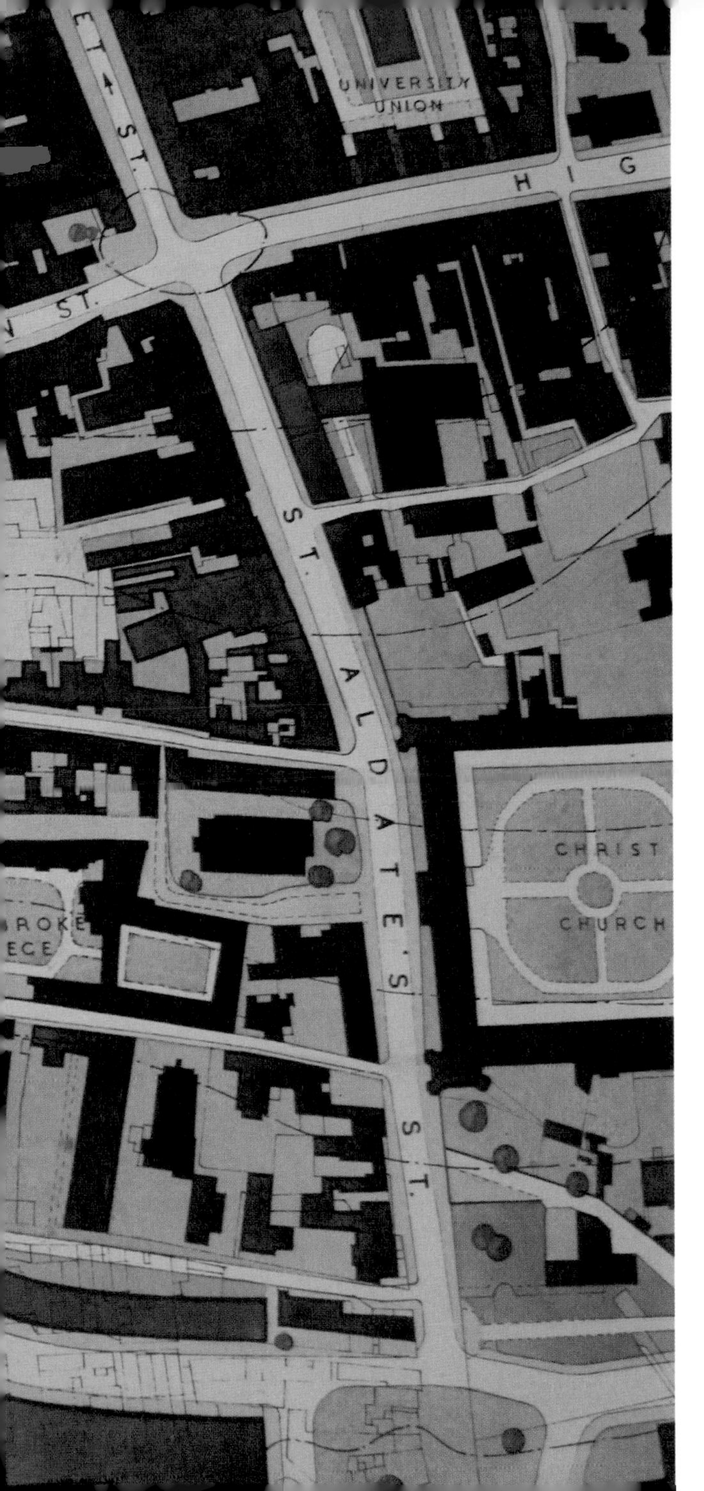

42

A NEW LOOK FOR OXFORD

Thomas Sharp was a prominent twentieth-century town planner, employed by Oxford City Council after the Second World War. In his daring vision, *Oxford Replanned*, Sharp devised a series of radical plans for the transformation of the city, which were never to materialize. This map features the central area west of Carfax, its new road network of graceful curves sweeping north from a roundabout near the police station to another interchange at Broad Street's western end, via a major junction at a location currently occupied by Bonn Square. There is also a grand building labelled 'Town Hall' close to the present County Hall.

Thomas Sharp, 'Central Oxford', 1948. (R) C17:70 Oxford (246)

Sharp produced another map to illustrate his plans to alleviate High Street's traffic congestion. This links St Aldate's and The Plain with a new road, Merton Mall, taking traffic out of the built-up area and across Christ Church Meadow instead. Merton Mall, which would have torn up the ground at the southern frontage of Christ Church, can be seen bisecting Broad Walk and dropping over the River Cherwell to a roundabout at The Plain, where the entrance to Cowley Road disappears beneath a landscaped park. The proposals failed to materialize but this planned relief road, which would have radically changed the city, was not finally rejected until the mid-1960s.

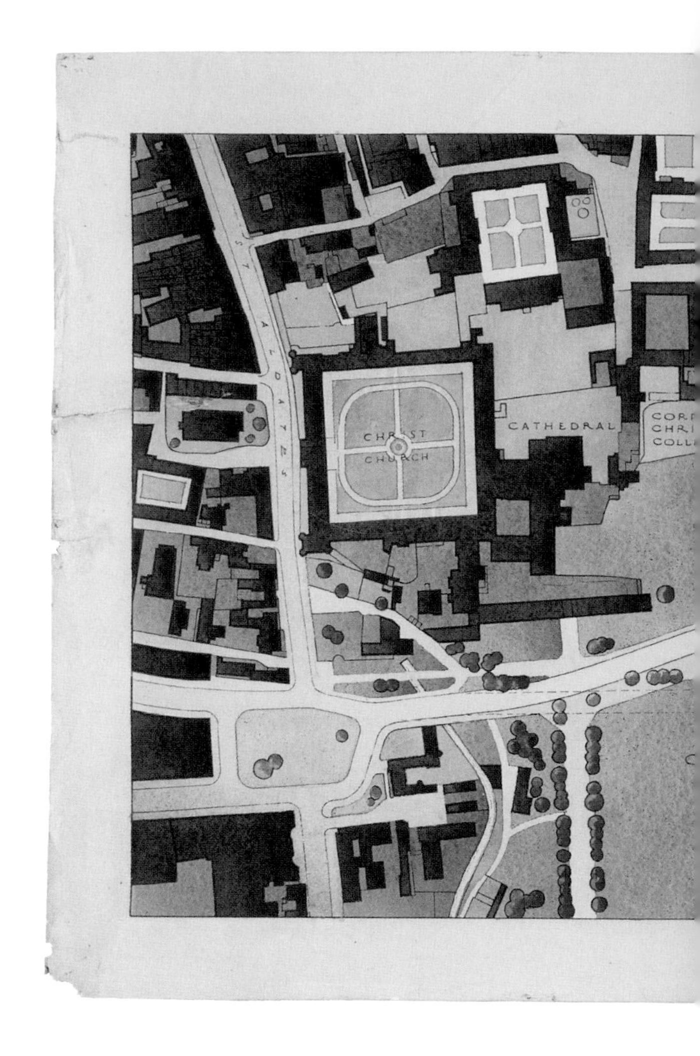

Thomas Sharp, 'Christ Church Meadow', 1948. (R) C17:70 Oxford (246)

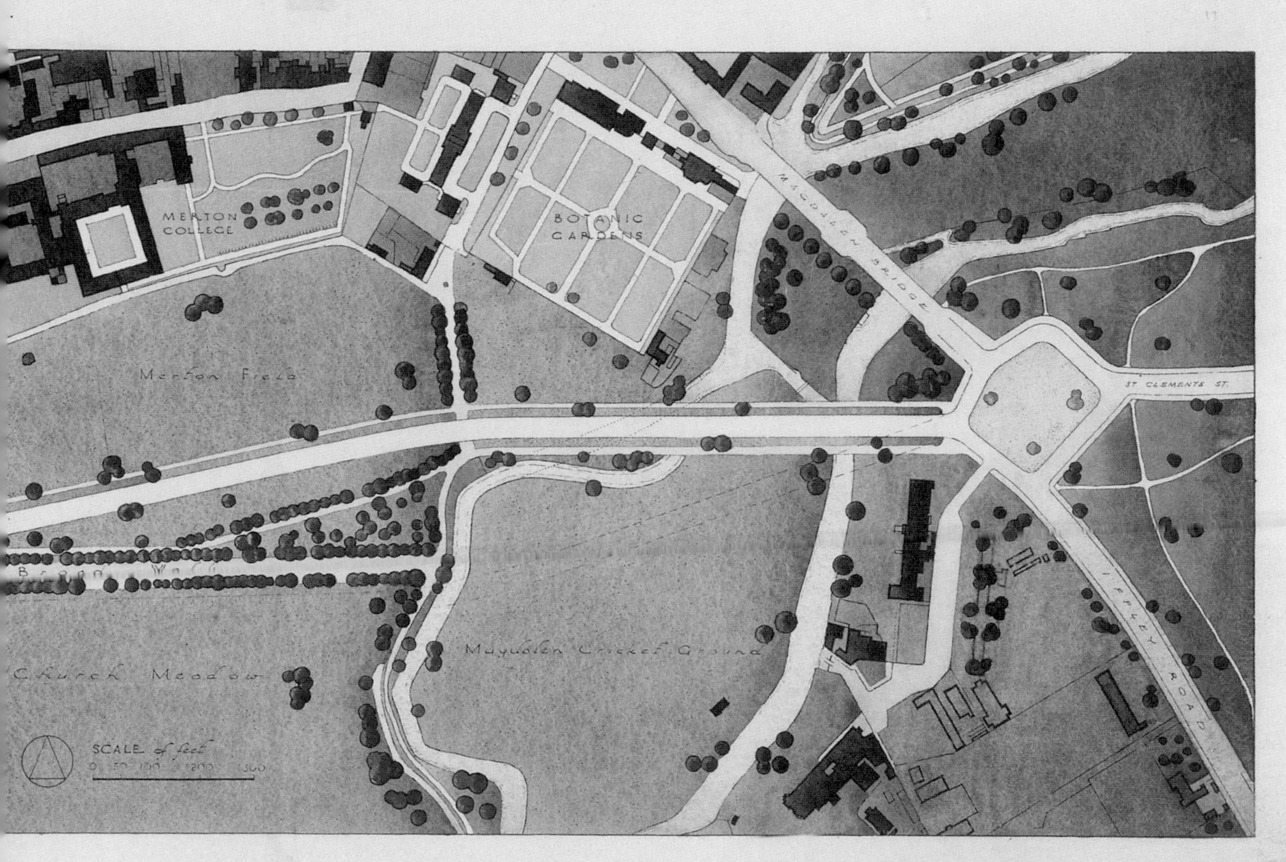
MERTON COLLEGE
BOTANIC GARDENS
MAGDALEN BRIDGE
ST CLEMENTS ST
Merton Field
Magdalen Cricket Ground
Church Meadow
IFFLEY ROAD
SCALE of feet
0 50 100 200 300

43

EXERCISE SURPRISE PACKET

In 1951 Exercise Surprise Packet – Britain's biggest military manoeuvres since the Second World War, involving 50,000 troops – used this map of imaginary geography. Most of England and Wales is labelled 'Anglian Peninsula (South)'. The 'peninsula' has two national boundaries creating Fantasia, Midland and Southland, but all towns and airfields featured are correctly located. The inset shows Fantasia joined to continental Europe (renamed 'Europa') via a land bridge towards Scandinavia. Along the Midland/Southland boundary lies a greater density of settlements with some unexpected additions: much of Essex is inundated; the 'Chequers Line' runs along the border; and two lakes have appeared in Wiltshire.

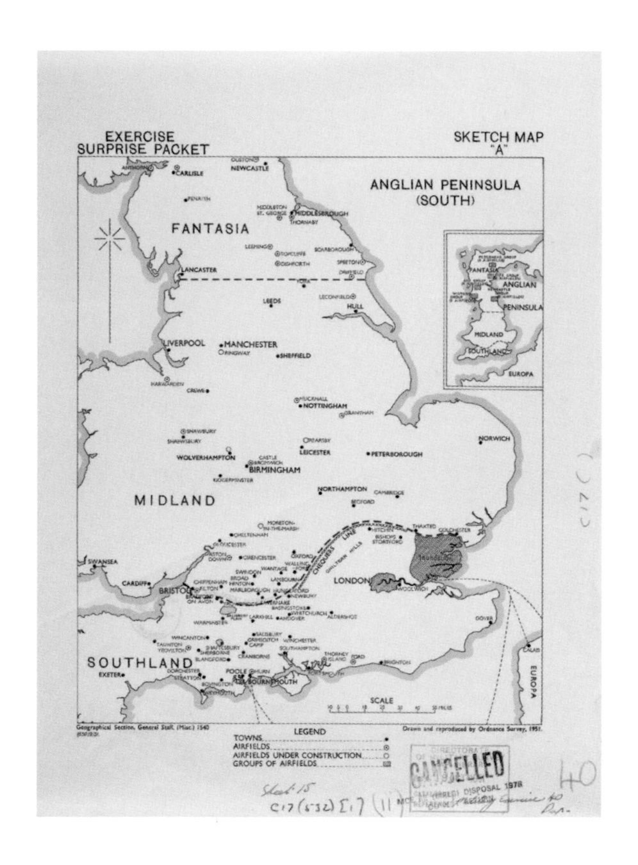

'Exercise Surprise Packet', Sketch Map 'A', 1951.
C17 (532) [1]

ANGLIAN PENINSULA (SOUTH)
FANTASIA
PENRITH
MIDDLETON ST. GEORGE
MIDDLESBROUGH
THORNABY
LEEMING
TOPCLIFFE
DISHFORTH
SCARBOROUGH
SPEETON
LANCASTER
DRIFFIELD
YORK
LECONFIELD
LEEDS
HULL
LIVERPOOL
MANCHESTER
RINGWAY
SHEFFIELD
HARWARDEN
CREWE
HUCKNALL
NOTTINGHAM
GRANTHAM
SHAWBURY
SHREWSBURY
REARSBY
LEICESTER
WOLVERHAMPTON
CASTLE BROMWICH
BIRMINGHAM
PETERBOROUGH
NORWICH
KIDDERMINSTER
NORTHAMPTON
CAMBRIDGE
BEDFORD
MIDLAND
MORETON-IN-THE-MARSH
CHELTENHAM
GLOUCESTER
HITCHIN
THAXTED
COLCHESTER
BISHOPS STORTFORD
CHEQUERS LINE
CHILTERN HILLS
Inundation
ASTON DOWN
CIRENCESTER
OXFORD
WALLING-FORD
WANTAGE
SWINDON
BROAD HINTON
LAMBOURN
CHIPPENHAM
FILTON
ARDIFF
BRISTOL
LONDON
WOOLWICH
MARLBOROUGH
HUNGERFORD
NEWBURY
BRADFORD ON AVON
AVON-KENNET CANAL
TROWBRIDGE
SAVERNAKE
BASINGSTOKE
SALISBURY PLAIN
LARKHILL
ANDOVER
WHITCHURCH
ALDERSHOT
DOVER
PETERHEAD GROUP (3 AIRFIELDS)
FANTASIA
FIFE GROUP (3 AIRFIELDS)
AYR GROUP (4 AIRFIELDS)
ANGLIAN
NEWCASTLE GROUP (5 AIRFIELDS)
WIGTOWN GROUP (5 AIRFIELDS)
PENINSULA
MIDLAND
SOUTHLAND
EUROPA

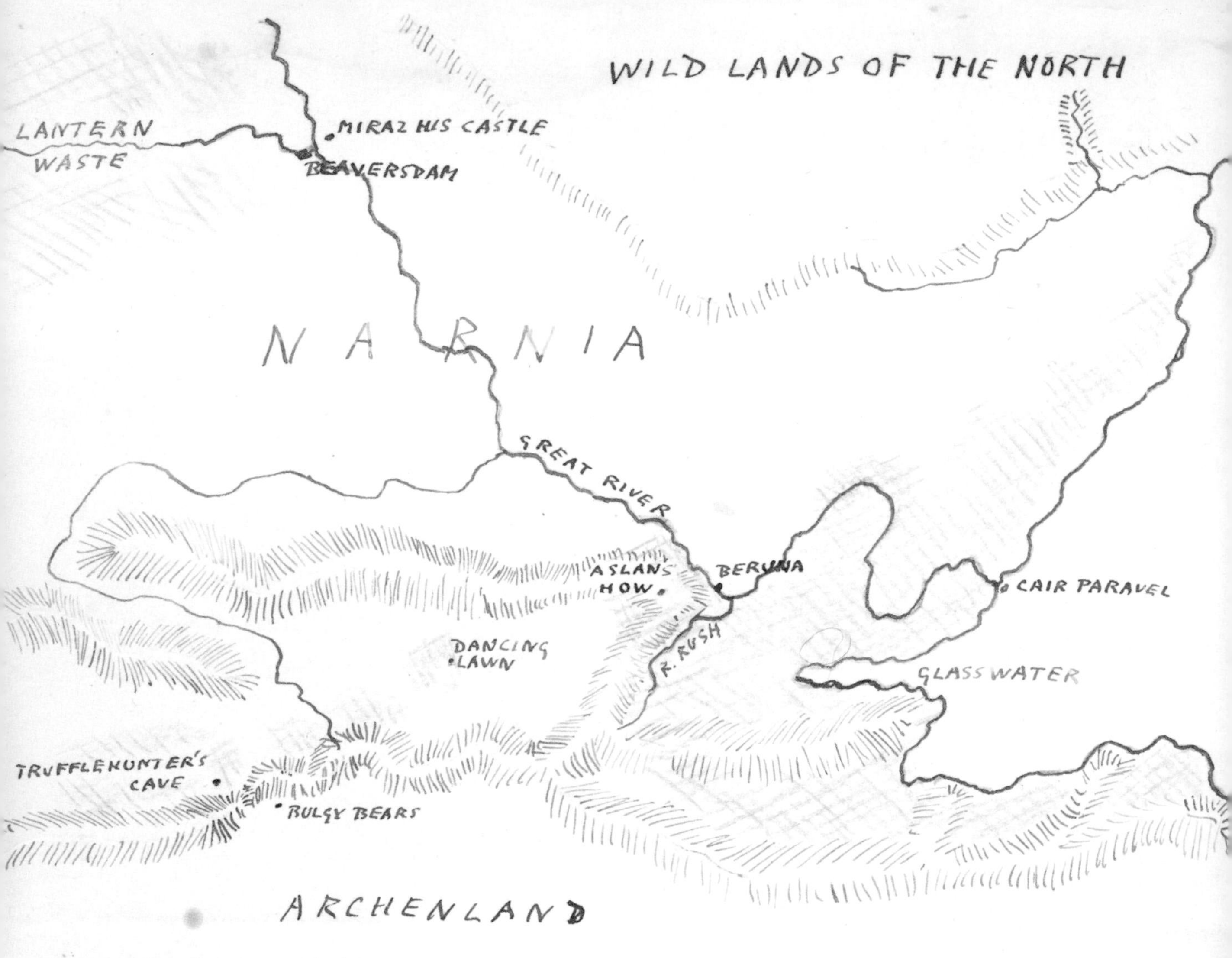

The ridge between Narnia and the Wild Lands of the North is only low hills: that between Narnia and Archenland, real mountains.

Aslan's How is on a moderate hill: but the range of which it is the Eastern end gets higher as it goes Westward.

Green = major woods.

A future story will require marshes here. We needn't mark them now, but must not put in anything inconsistent with them!

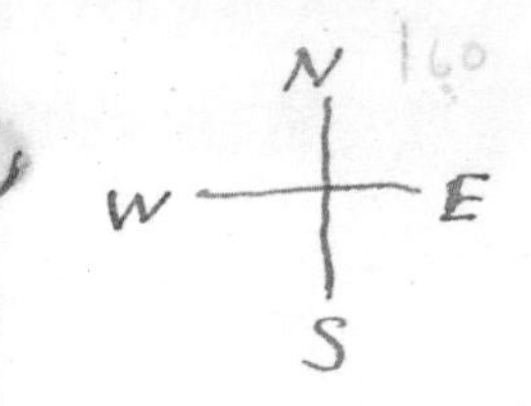

44

C.S. LEWIS'S MAP OF NARNIA

When C.S. Lewis began writing *The Chronicles of Narnia* in the 1950s, he turned to the artist Pauline Baynes to draw a map of his fictional landscape. He drew this one by hand as a guide for Baynes's subsequent maps, which he told her should be more 'like a medieval map than an Ordnance Survey – mountains and castles drawn'. This sketch provides key locations for his stories: 'Lantern Waste', the first Narnian location described in the *Chronicles*, where Lucy Pevensie meets Mr Tumnus; 'Cair Paravel', Narnia's capital; and the 'Wild Lands of the North', where the White Witch was exiled.

C.S. Lewis, 'Map of Narnia for Prince Caspian', 1951. MS. Eng. lett. C.220/1, fol. 160

45

SOVIET MAP OF THE MEDWAY TOWNS

Mapped at a scale of 1:10,000, this two-sheet town plan is lavishly printed in ten colours and was created by Soviet military cartographers in 1984. All streets are included and named, as are principal buildings. The amount of geographical information is huge. Rochester Bridge is annotated thus: ЖБ 6 300-15 / сб. 100. This indicates it is built of reinforced concrete, with a 6m clearance above the water; the bridge is 300m long by 15m wide, with a load-bearing capacity of 100 tonnes. This map was probably created using 1930s Ordnance Survey maps alongside up-to-date intelligence 'on the ground'. Soviet town mapping was a clandestine global project that only came to the West's attention after the breakup of the USSR.

Chatem, Djillingem, Rochester, 1984. C17:70 Chatham (4)

ГЕНЕРАЛЬНЫЙ ШТАБ

ЧАТЕМ, ДЖИЛЛИНГЕМ, РОЧЕСТЕР

(M-31-13,14)

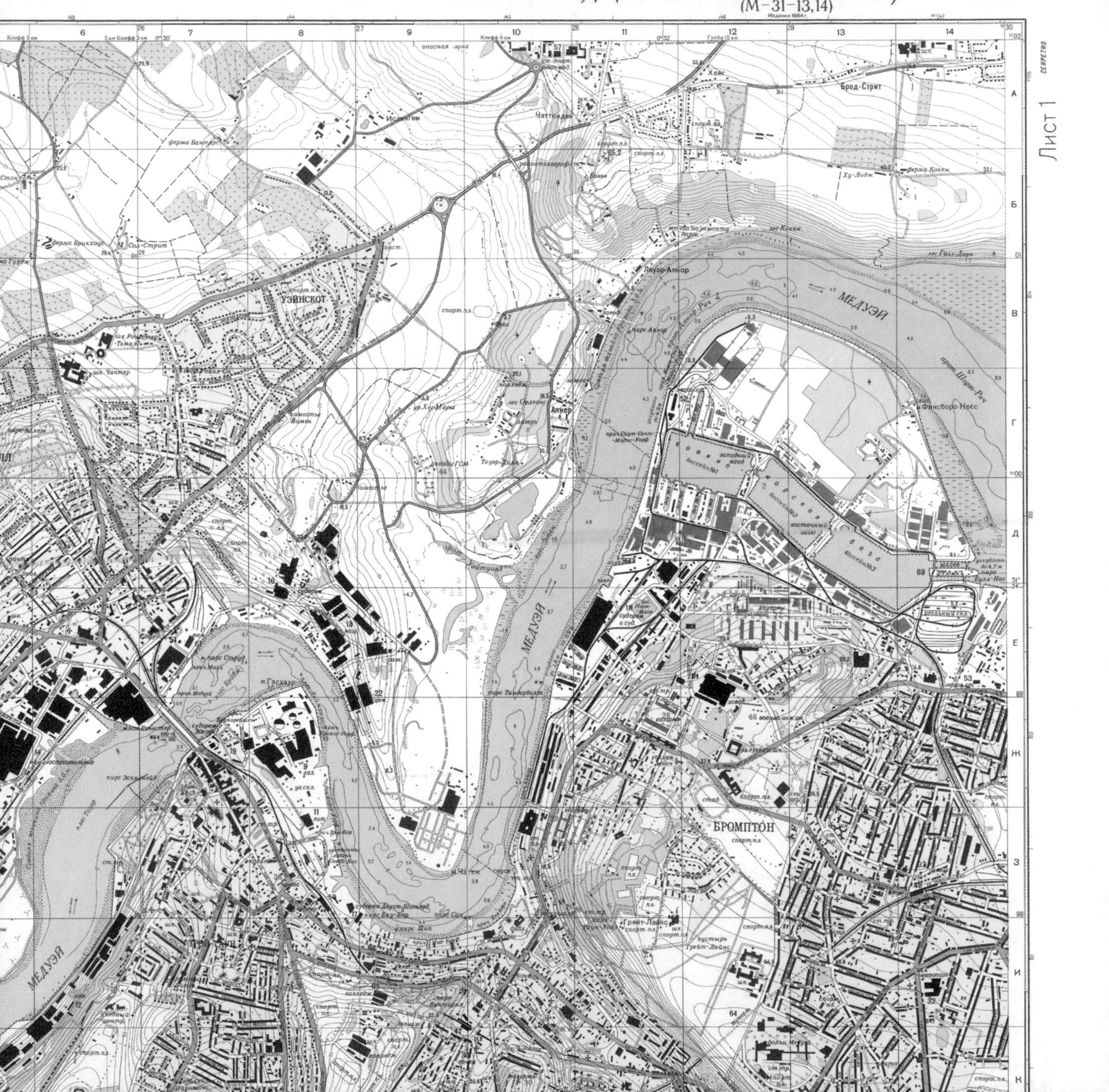

46

TYNE AND WEAR – OR IS IT?

A topographic map of Tyne and Wear, or not? Layla Curtis's collage takes extracts from American and Australian maps, added to some British, Canadian, Irish, Jamaican and New Zealand cartography. Segments are extracted and reassembled to create a reassuringly familiar cartographic landscape, albeit one very different to the region's physical geography. The coastline is close to our expectations, as are the Tyne and Wear estuaries. There are, however, fifty-two Newcastles and ten North Easts, plus references to the region's former economic driver, Coal, and Shearer Point, in honour of Newcastle United's record goal-scorer. Illustrating the map's artistic preoccupations, 'Baltics Corners' is placed precisely where we expect to find Gateshead's Baltic Art Gallery.

Layla Curtis, *NewcastleGateshead*, 2005. O1 (30)

NORTHUMBERLAND
DUDLEY
KARLANTIJPA NORTH
DESERT
NEWCASTLE
Newcastle Waters
WILSON
Shields
Newcastle
Raymond Terrace
Newcastle Airport
Beachlands Estate
NORTHUMBERLAND STRAIT
GATESHEAD ISLAND
CB-29/67F
North East I.
Hexham I.
Berwick I.
Carlisle I.
Cape Northumberland
Louth Bay
North Shields
Port Hunter
Tyneham
Broad Bench
NORTH EAST
North East
North East Harbour
Northeast Cove
North East Heights
Walkers Corners
Fellingsbro
Örnbrotorp
SUNDERLAND
Sunderland
Pinedale
Houghton
WASHINGTON
Washington
Washington Heights
Middlesboro
WHITBY
Cleveland
Port Whitby
DÉTROIT DE NORTHUMBERLAND

47

GRAYSON PERRY'S *MAP OF NOWHERE*

The artist Grayson Perry has always been fascinated by the imaginative possibilities of mapping, especially those derived from its medieval history. His 'Map of Nowhere' puns on the Renaissance idea of mapping Utopia (which can mean 'nowhere'), as well as drawing explicitly on medieval *mappae mundi* that show the world embodied as Christ. In this etching Perry is Christ, a playful reflection on our modern obsession with the self as the centre of everything. Instead of the theological certainties of medieval Christianity, Perry's world teems with modern secular institutions and concerns, from Starbucks and Microsoft to 'Despair' and 'Insecurity'.

NOWHERE
Love
Peace
Beauty
Truth
Despair
Saint Claire

False Hope
Delusions of Grandeur
Inhuman Logic
Rationalism
Plausible reality
The Limits of Reason
BLOOD
Time
HERE AND NOW
Choice without consequences
Experience
Thumb Generation
Stuff happens
Merchandising
Faith-based intelligence
selfish -gene
Trivia
The New Black
Cool-hunter
Kidults
Shopaholic
BoBos
Regrets
Disappointment
Everything changes
Infantalism
The Sadness of the Excessively Logical
Despair
A cure for Evil
Wisdom
Uncertainty
Doubt
Preachy
Mortality
Saint Claire
Lazy
Angry
Unfulfilled
Blaming
Sexism
Lonely
Self-delusion
False bonhomie
Latest must-have
Trendy Wine Bar
Hubris
Logo board
A Defining Enemy
Know-all
Internet dating

At the centre of Perry's map stands the island of 'Doubt'. Where medieval Christian maps would show Jerusalem and the hope of salvation through Christ's crucifixion, Perry captures our modern secular doubts and fears concerning any kind of personal redemption. Grand political or theological belief systems literally lead 'nowhere'. 'Doubt' is surrounded by 'Uncertainty', 'Hubris' and the 'Meaningless' sea. To undercut any assumptions that he is some kind of prophet, Perry creates a visual joke at his own expense, showing pilgrims flocking to a monastery illuminated by light coming from his rectum. His is a map that consoles with no certainties.

Previous page Grayson Perry, *Map of Nowhere*, 2008. Details of the island of Doubt (opposite page) and the pilgrims (above)

48

UK ELECTIONS 2017

These representations of the 2017 British general election results demonstrate the differences between conventional projections of land area ('Geographic view'); cartograms where each constituency is represented by a hexagonal area ('Constituency view') and gridded population cartograms, where each area is proportional to its population ('Population view'). Each of these representations allows for a different interpretation of the information visualized, thus broadening our perspective on how the political landscape of the country can be understood. Conventional maps over-represent less densely populated rural areas that are often also very different in their electoral geography from more populated areas.

Map representations of the 2017 general election results in the United Kingdom by Benjamin Hennig, 2017

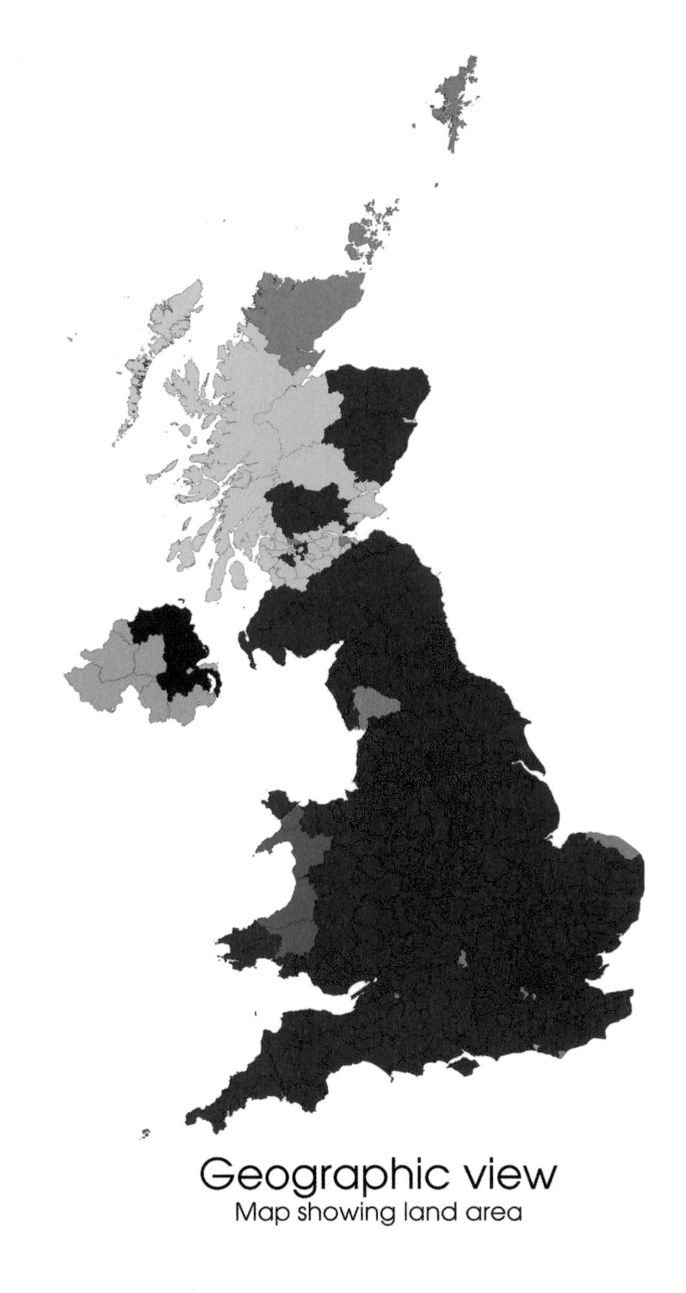

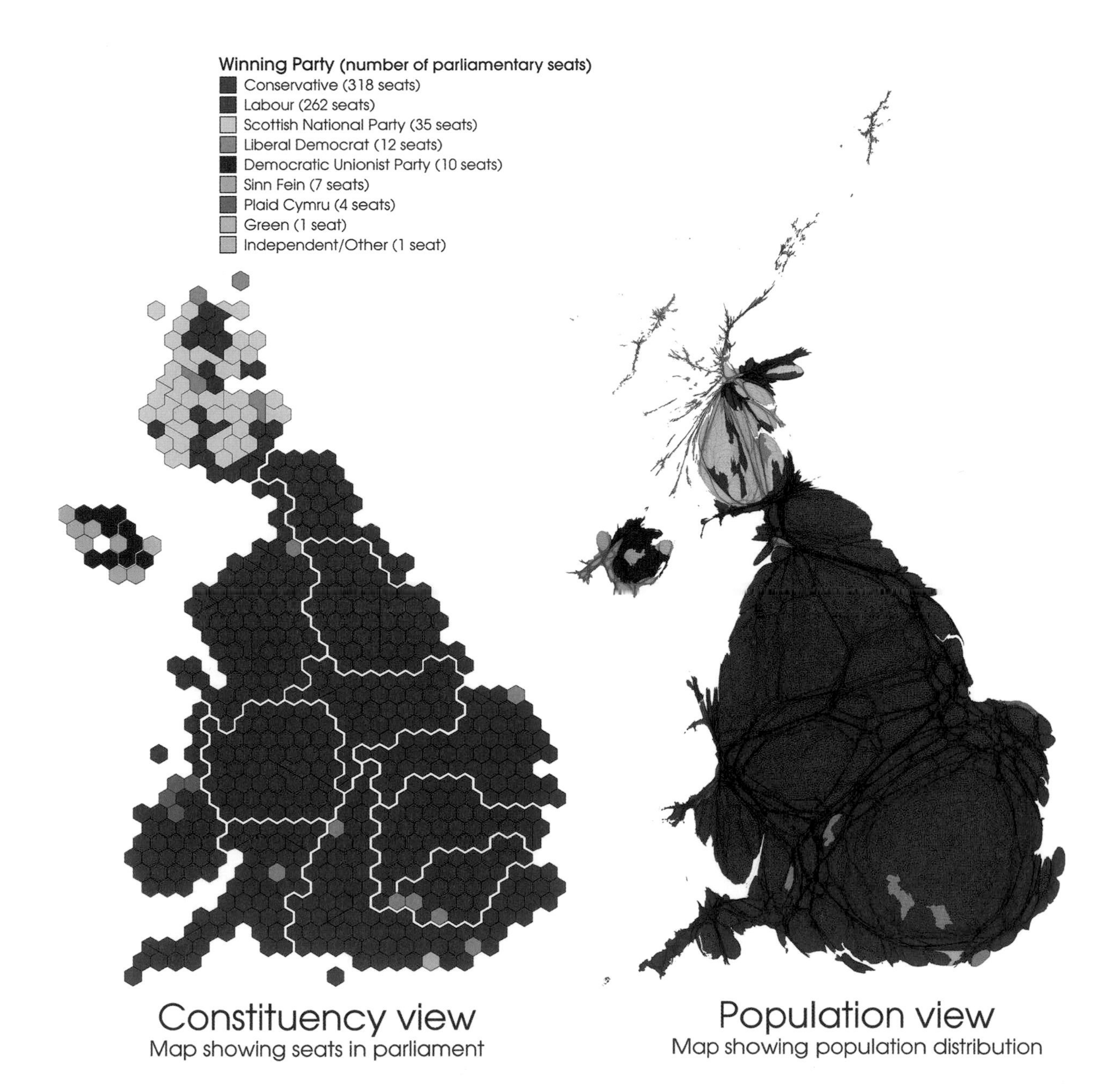

Constituency view
Map showing seats in parliament

Population view
Map showing population distribution

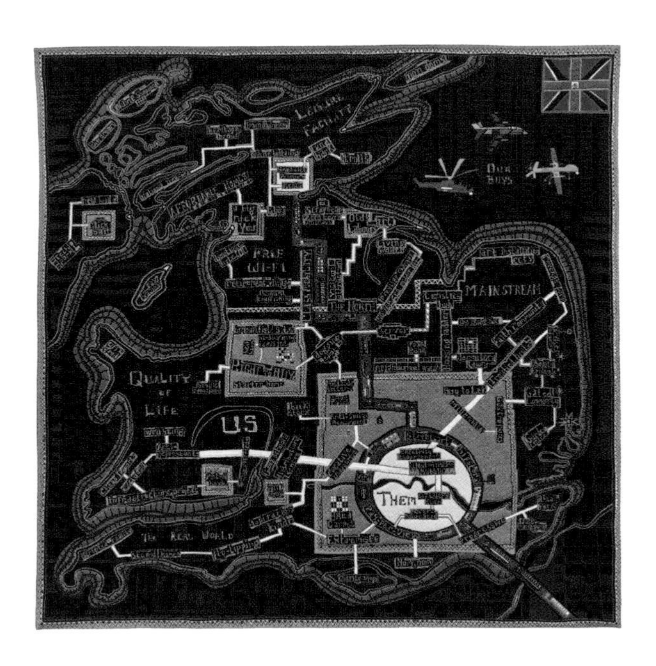

49

GRAYSON PERRY'S *THE RED CARPET*

In 2017 Grayson Perry designed this tapestry map of the United Kingdom, partly in response to the country's 'Brexit' vote to leave the European Union in June 2016. This was hardly the welcoming 'red carpet' of a nation at ease with itself. Perry described it as 'a map of British society' reflecting 'the density of population rather than the lie of the land'. Scotland is tiny in contrast to London. The country is covered in 'buzzphrases' that Perry believed captured the national mood: 'Millenials' [sic], 'Gentrification' and 'Zero Hours Contract'. It expresses the state of the nation at a crucial moment in its history.

Grayson Perry, *The Red Carpet*, 2017

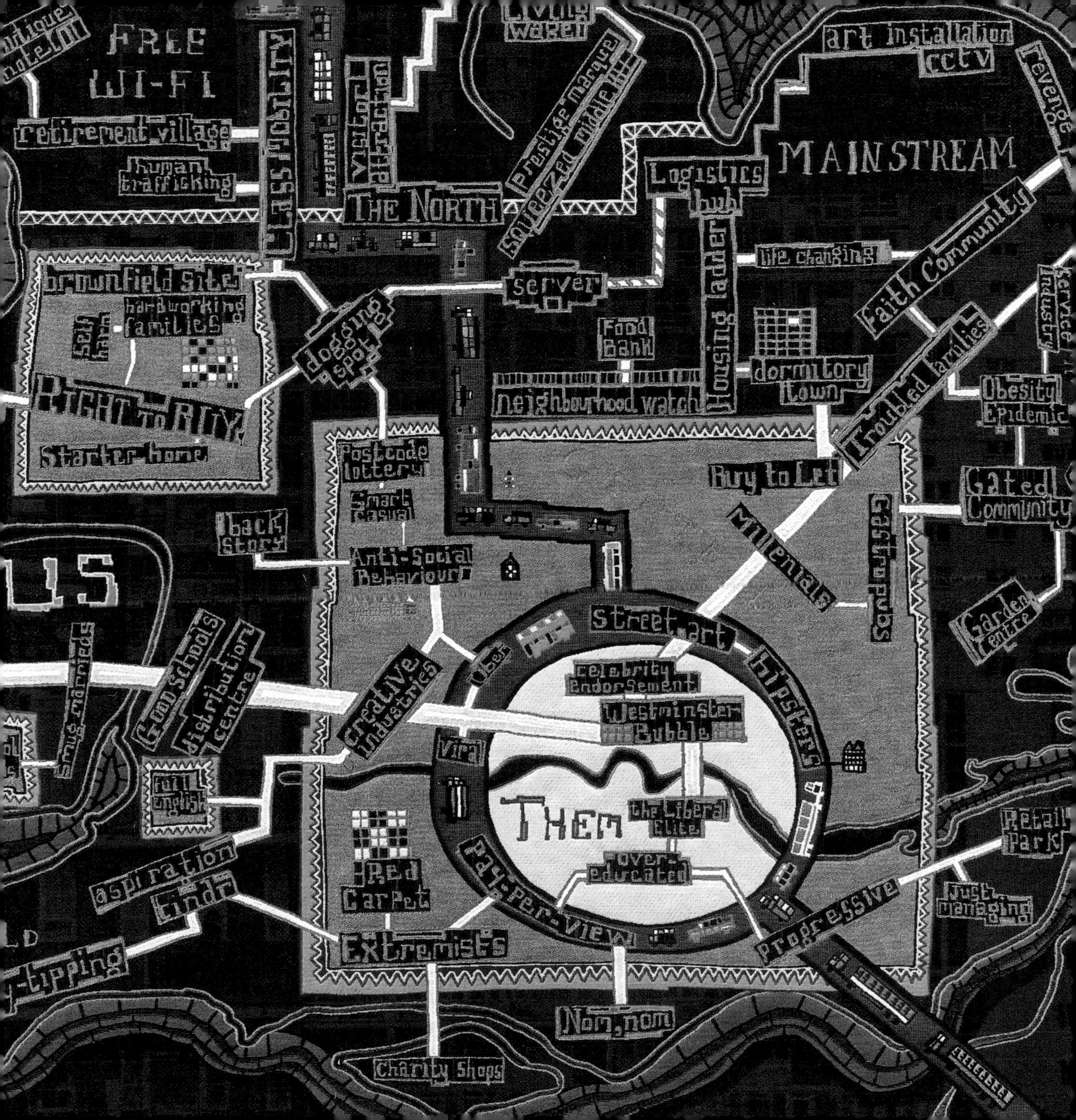

FREE WI-FI
retirement village
human trafficking
CLASS MOBILITY
visitor attraction
THE NORTH
prestige marque
squeezed middle
Logistics hub
MAINSTREAM
art installation
cctv
revenge
life changing
Faith Community
server
Food Bank
Housing ladder
dormitory town
neighbourhood watch
Troubled families
Service Industry
Obesity Epidemic
brownfield site
hardworking families
self harm
dogging spot
RIGHT to BUY
Starter home
Postcode lottery
Smart casual
back story
Anti-Social Behaviour
Buy to Let
millenials
Gastropubs
Gated Community
Garden Centre
Street art
hipsters
celebrity endorsement
Westminster Bubble
Uber
smug marrieds
Good schools
distribution centre
creative industries
viral
THEM
the Liberal Elite
over-educated
Full English
aspiration
Red Carpet
Extremists
Pay-Per-View
Progressive
Retail Park
Just managing
Nom, nom
Charity Shops

50

MAPPA MUNDI-STYLE POPULATION CARTOGRAM

Digital cartography and computational statistics are transforming the possibilities of data analysis and their visualization, generating new ways of showing the increasing complexity of a globalized world. An extreme and conceptual modification of a gridded approach to cartograms is reflected in this modern interpretation of a medieval *mappa mundi*, imaging a world without its (unpopulated) oceans. India and Asia are shown at the world's centre, recognizing the most populated region of the planet, with the Americas and Europe cast aside to the periphery as a consequence of their relatively low populations. Cartograms have become viable alternatives to conventional maps by presenting and understanding geographical data in novel ways.

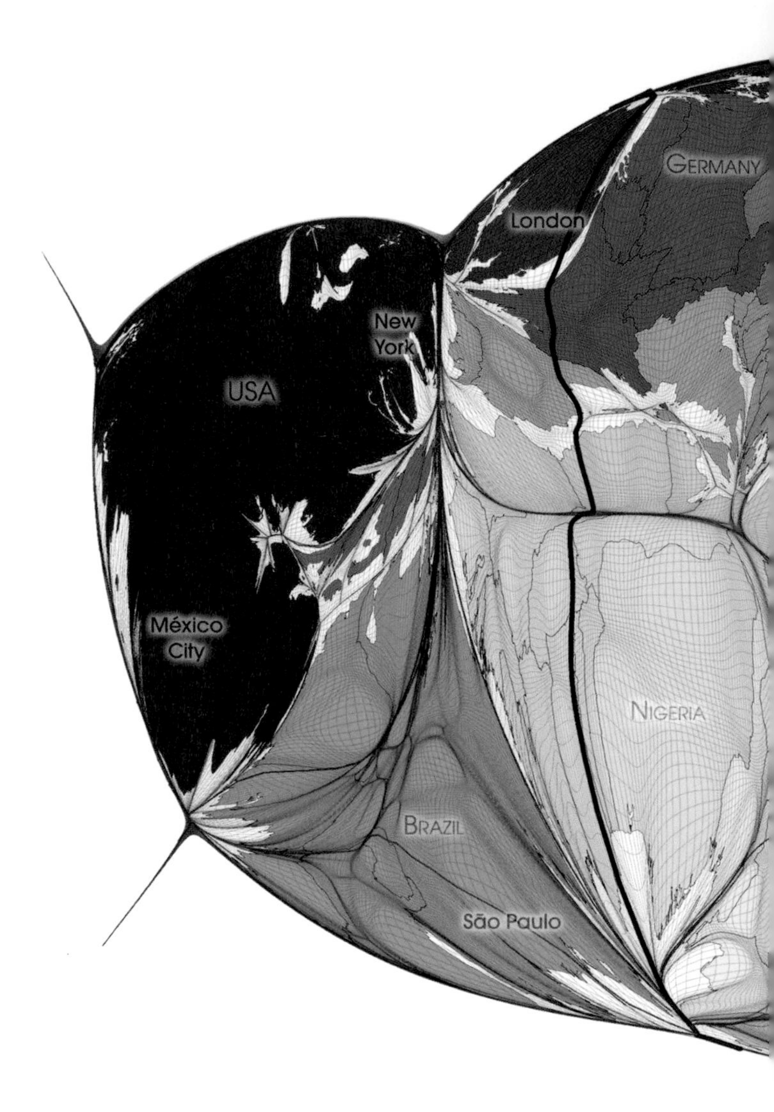

Mappa mundi-style gridded population cartogram of the world by Benjamin Hennig, 2018

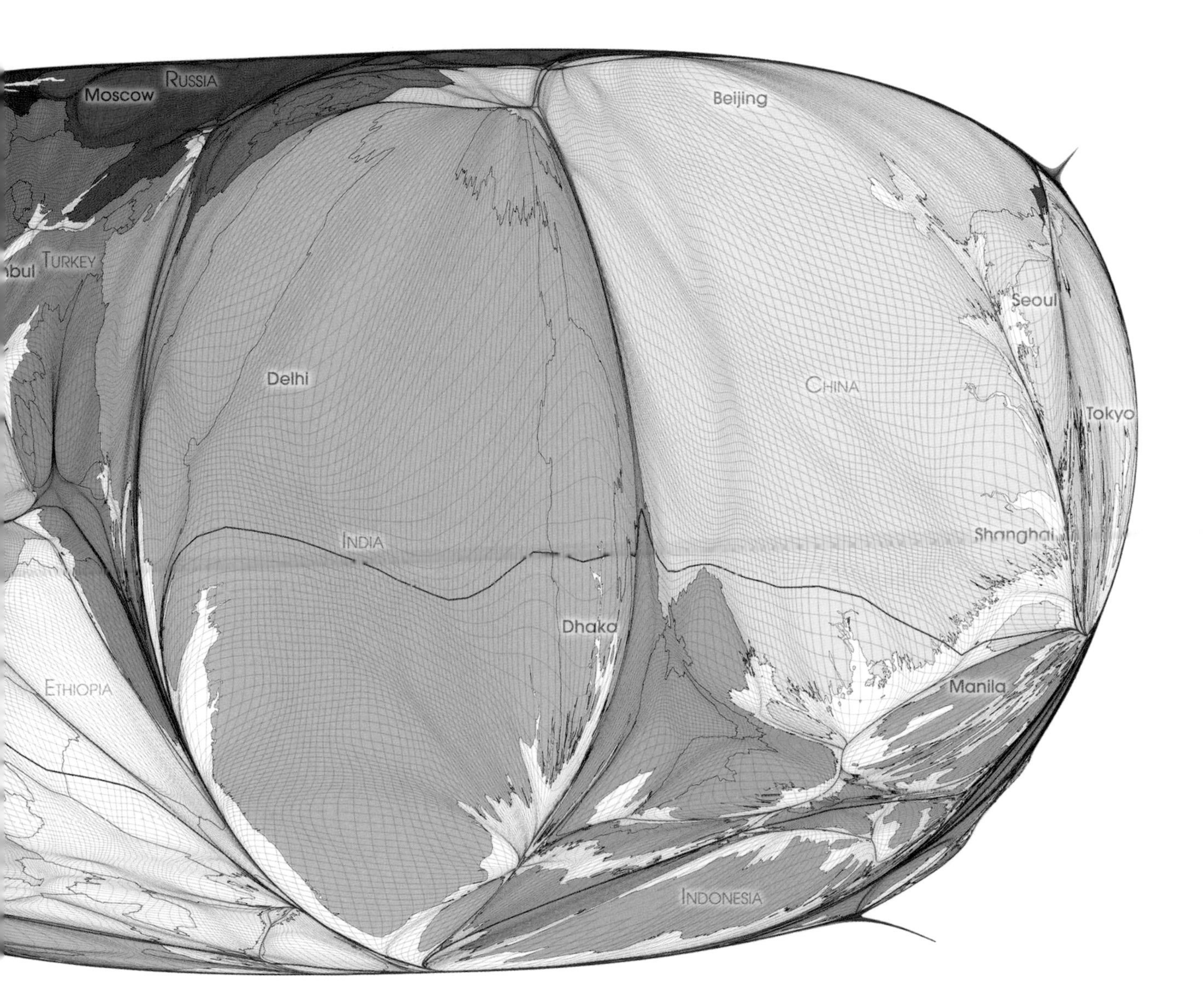

Russia
Moscow
Beijing
Turkey
nbul
Seoul
Delhi
China
Tokyo
India
Shanghai
Dhaka
Ethiopia
Manila
Indonesia

FURTHER READING

Barber, P., *The Map Book*, Weidenfeld & Nicolson, London, 2005

Beckett, J.V., *A History of Laxton: England's Last Open-Field Village*, Basil Blackwell, Oxford, 1989

Birkholz, D., *The King's Two Maps: Cartography and Culture in Thirteenth-Century England*, Routledge, London, 2004

Board, C., 'Air Photo Mosaics: A Short-Term Solution to Topographic Map Revision in Great Britain 1944–51', *Sheetlines*, 71, 2004, pp. 24–35

Brotton, J., *A History of the World in Twelve Maps*, Penguin, London, 2012

Chasseaud, P., *Topography of Armageddon: A British Trench Map Atlas of the Western Front 1914–1918*, Mapbooks, Lewes, 1991

Chubb, T., *A Descriptive Catalogue of the Printed Maps of Gloucestershire 1577–1911*, J.W. Arrowsmith, [Bristol], 1913

Crane, N., *Mercator: The Man who Mapped the Planet*, Phoenix, London, 2003

Darkes, G. and Spence, M., *Cartography: An Introduction*, 2nd edn, British Cartographic Society, London, 2017

Davies, J. and Kent, A., *The Red Atlas: How the Soviet Union Secretly Mapped the World*, University of Chicago Press, Chicago and London, 2017

Delano-Smith, C., et al., 'New Light on the Medieval Gough Map of Britain', *Imago Mundi*, 69 (1), 2017, pp. 1–36

Dorling, D., 'New Maps of the World, its People, and their Lives', *Society of Cartographers Bulletin*, 39 (1/2), 2006, pp. 35–40

Gastner, M.T. and Newman, M.E.J., 'Diffusion-Based Method for Producing Density Equalizing Maps', *Proceedings of the National Academy of Sciences USA*, 101, 2004, pp. 7499–504

Hall, D., *Treasures from the Map Room: A Journey through the Bodleian Collections*, Bodleian Library, Oxford, 2016

Harley, J.B. and Woodward, D. (eds), *History of Cartography,* University of Chicago Press, Chicago and London, 1987–

Hennig, B.D., 'The Human Planet', *Environment & Planning A*, 45 (3), 2013, pp. 489–91

Hennig, B.D., *Rediscovering the World: Map Transformations of Human and Physical Space*, Springer, Berlin, 2013

Hennig, B.D., 'Worldmapper: Rediscovering the World', *Teaching Geography*, 43 (2), 2018, pp. 66–8

MacCannell, D., *Oxford: Mapping the City*, Birlinn, Edinburgh, 2016

Millea, N., *The Gough Map: The Earliest Road Map of Great Britain?*, Bodleian Library, Oxford, 2007

Monmonier, M., *Rhumb Lines and Map Wars: A Social History of the Mercator Projection*, University of Chicago Press, Chicago and London, 2004

Orwin, C.S. and C.S., *The Open Fields*, 3rd edn, Oxford University Press, Oxford, 1967

Rapoport, Y. and Savage-Smith, E. (eds), *An Eleventh-Century Egyptian Guide to the Universe: The Book of Curiosities*, Brill, Leiden, 2014

Rapoport, Y., *Maps of Islam*, Bodleian Library, Oxford, forthcoming

Sharp, T., *Oxford Replanned*, Architectural Press, London, 1948

Sharp, T., *Oxford Replanned: Exhibition of the Planning Proposals made by Thomas Sharp*, Oxford University Press, Oxford, 1948

Tobler, W.R., 'Thirty-Five Years of Computer Cartograms', *Annals of the Association of American Geographers*, 94 (1), 2004, pp. 58–73

Tooley, R.V., *Maps and Map-Makers*, 7th edn, Batsford, London, 1987

Turner, H.L., *No Mean Prospect: Ralph Sheldon's Tapestry Maps*, Plotwood Press, [Derby], 2010

Turner, H.L., 'The Sheldon Tapestry Maps belonging to the Bodleian', *Bodleian Library Record*, 17 (5), 2002, pp. 293–313

Turner, H.L., 'The Sheldon Tapestry Maps: Their Content and Context', *Cartographic Journal*, 40 (1), 2003, pp. 39–49

Tyacke, S. and Huddy, J., *Christopher Saxton and Tudor Map-Making*, British Library, London, 1980

Wallis, H.M. and Robinson, A.H., *Cartographical Innovations: An International Handbook of Mapping Terms to 1900*, Map Collector Publications/ International Cartographic Association, [Tring], 1987

Wells-Cole, A., 'The Elizabethan Sheldon Tapestry Maps', *Burlington Magazine*, 132 (1047), 1990, pp. 392–401

Worldmapper, 2018, *Rediscover the World as You've Never Seen it Before*, www.worldmapper.org

ACKNOWLEDGEMENTS

This book grew out of the *Talking Maps* exhibition held at the Bodleian Library from 5 July 2019 to 1 March 2020.

We are very grateful to the following institutions and individuals for permission to reproduce materials: the Pitt Rivers Museum, the Ashmolean Museum, Oxford City Council and Grayson Perry. Various individuals have generously shared their time and expertise in answering our questions and reading material. We would like to thank Benjamin Hennig for joining the project towards the end and enriching it enormously; Alfred Hiatt for helping with various medieval cartographic queries; Keith Lilley for providing crucial help on the Gough map; Adam Lowe and his team at Factum Arte for their ground-breaking digital work on various maps; Michael Athanson for his cartographic input on the Laxton map; Stuart Ackland, Debbie Hall and Tessa Rose from the Bodleian Maps team for their help in identifying and delivering maps for the volume; Stephen Johnston at the History of Science Museum, University of Oxford, for his expertise on astrolabes and compasses; Charles Manson for crucial assistance and extensive feedback on the Tibetan material, Catherine McIlwaine on Tolkien and Pnina Arad on pilgrimage and William Wey; Jack Langton for examining the Laxton map with us; and Yossef Rapoport for sharing his unparalleled knowledge of early Islamic mapping, including unpublished work. The Bodleian Library publishing team have been a model of professionalism and patience, and a joy to work with in producing such a beautiful book: thank you Samuel Fanous, Janet Phillips and Leanda Shrimpton. We have also been indebted to the Bodleian exhibitions team, especially Madeline Slaven and Sallyanne Gilchrist for their guidance and eye for detail when planning the *Talking Maps* exhibition. Nick would like to dedicate the book to Alice and Evvie; Jerry would like to dedicate it to Ruby, Hardie and Honey.

PICTURE CREDITS

Where no name is given in the caption, the creator of the map is unknown.

17 Image courtesy of Daniel Crouch Rare Books – crouchrarebooks.com; 28, 32 © Pitt Rivers Museum, University of Oxford; 37 Photo © Oxford University Images / Oxfordshire History Centre; 37 Map © Oxford City Council; 33 Ashmolean Museum, University of Oxford; 38 © The Tolkien Estate Ltd, 1986; 38 © The Tolkien Estate Ltd; 44 Copyright © C.S. Lewis Pte. Ltd. 1950. Reprinted by permission; 46 Reproduced courtesy of Layla Curtis; 47, 49 Grayson Perry and Paragon | Contemporary Editions Ltd, London; 48, 50 © Ben Hennig / Worldmapper.org.

INDEX

Numbers of pages with illustrations are in *italics*